NEUROCOGNITIVE AND PHYSIOLOGICAL FACTORS DURING HIGH-TEMPO OPERATIONS

Human Factors in Defence

Series Editors:

Dr Don Harris, Managing Director of HFI Solutions Ltd, UK
Professor Neville Stanton, Chair in Human Factors of Transport at the University of Southampton, UK
Dr Eduardo Salas, University of Central Florida, USA

Human factors is key to enabling today's armed forces to implement their vision to 'produce battle-winning people and equipment that are fit for the challenge of today, ready for the tasks of tomorrow and capable of building for the future' (source: UK MoD). Modern armed forces fulfil a wider variety of roles than ever before. In addition to defending sovereign territory and prosecuting armed conflicts, military personnel are engaged in homeland defence and in undertaking peacekeeping operations and delivering humanitarian aid right across the world. This requires top class personnel, trained to the highest standards in the use of first class equipment. The military has long recognised that good human factors is essential if these aims are to be achieved.

The defence sector is far and away the largest employer of human factors personnel across the globe and is the largest funder of basic and applied research. Much of this research is applicable to a wide audience, not just the military; this series aims to give readers access to some of this high quality work.

Ashgate's *Human Factors in Defence* series comprises of specially commissioned books from internationally recognised experts in the field. They provide in-depth, authoritative accounts of key human factors issues being addressed by the defence industry across the world.

Neurocognitive and Physiological Factors During High-Tempo Operations

EDITED BY

STEVEN KORNGUTH
REBECCA STEINBERG
University of Texas at Austin, USA

&

MICHAEL D. MATTHEWS
United States Military Academy, USA

ASHGATE

Published by
Ashgate Publishing Limited
Wey Court East
Union Road
Farnham
Surrey, GU9 7PT
England

Ashgate Publishing Company
Suite 420
101 Cherry Street
Burlington
VT 05401-4405
USA

www.ashgate.com

British Library Cataloguing in Publication Data
Neurocognitive and physiological factors during high-tempo operations. -- (Human factors in defence)
1. Psychology, Military. 2. Soldiers--Job stress--United States. 3. Combat--Physiological aspects. 4. Cognition--Physiological aspects.
I. Series II. Kornguth, Steven E. III. Steinberg, Rebecca.
IV. Matthews, Michael D.
355.4'019-dc22

ISBN: 978-0-7546-7923-3 (hbk)
ISBN: 978-0-7546-9825-8 (ebk)

Library of Congress Cataloging-in-Publication Data
Neurocognitive and physiological factors during high-tempo operations / [edited] by Steven Kornguth, Rebecca Steinberg, and Michael D. Matthews.
p. cm.
Includes bibliographical references and index.
ISBN 978-0-7546-7923-3 (hardback) -- ISBN 978-0-7546-9825-8 (ebook) 1. Psychology, Military. 2. Soldiers--Job stress--United States. 3. Combat--Psychological aspects. 4. Combat--Physiological aspects. 5. Cognition--Physiological aspects. I. Kornguth, Steven E. II. Steinberg, Rebecca. III. Matthews, Michael D.
U22.3.N48 2010
355.401'9--dc22

2009046926

Mixed Sources
Product group from well-managed forests and other controlled sources
www.fsc.org Cert no. SA-COC-1565
© 1996 Forest Stewardship Council
FSC

Printed and bound in Great Britain by
MPG Books Group, UK

Contents

List of Figures

List of Tables

List of Abbreviations

5-HTT	Serotonin Transporter
AFNI	Analysis of Functional Neuroimages
ANT	Attentional Network Task
AOI	Area of Interest
APD	Avalanche Photodiode
API	Application Programming Interface
ARCIC	Army Capabilities Research Center, pronounced "R-Kick"
ARGOS	Automobile for Research in Ergonomics and Neuroscience
ASD	Acute Stress Disorder
BDNF	Bone Derived Neurotrophic Factor
BFT	Blue Force Tracker
BOLD	Blood Oxygen Level Dependent
CDA	Commander's Digital Assistant
CD-RISC	Connor-Davidson Resilience Scale
CFT	Cubic Feet
COL	Colonel
COMT	Catechol-O-Methyltransferase
COTS	Commercial Off-the-Shelf
CPOF	Command Post of the Future
CROP	Common Relevant Operating Pictures
CT	Computed Tomography
CW	Continuous Wave
DARPA	Defense Advance Research Projects Agency
DBN	Dynamic Bayesian Network
DES	Discrete Event Simulation
DHEA	Dehydroepiandosterone
DLPFC	Dorsolateral Prefrontal Cortex
DMS	Delayed Match-to-Sample
DoD	Department of Defense
DOT	Diffuse Optical Tomography
DSM IV	Diagnostic and Statistical Manual of Mental Disorders, 4th Edition
DTI	Diffusion Tensor Imaging
EEG	Electroencephalography
EKG	Electrocardiography
EMG	Electromyography
EPI	echo-planar imaging
ERP	Event-Related Potential

ESM	Experience Sampling Method
FA	Fractional Anisotropy
FBCB2	Force XXI Battle Command, Brigade-and-Below
FFW	Future Force Warrior
FIFO	First-In-First-Out
fMRI	Functional Magnetic Resonance Imaging
FOV	Field of View
FSO	Full Spectrum Operations
GDT	Game of Dice Task
GEN	General
GIS	Geographic Information System
GM	Grey Matter
GPS	Global Positioning System
HAF	Highest Attribute First
HALE	High Altitude Long Endurance
HMD	Helmet Mounted Display
HPE	Human Performance Enhancement
HPO	Human Performance Optimization
hrf	Hemodynamic Response Function
HTL	Human in the Loop
IBA	Interceptor Body Armor
ICA	Independent Components Analysis
ICC	Intraclass Correlation
ICG	Indocyanin Green
ID	Integrative Decision-Making
IED	Improvised Explosive Device
JIIM	Joint, Interagency, Intergovernmental and Multi-national
JMRC	Joint Multinational Readiness Center
JRTC	Joint Readiness Training Center
LED	Light Emitting Diode
LHPA	Limbic-Hypothalamo-Pituitary-Adrenal
Lt Cmdr	Lieutenant Commander
LTC	Lieutenant Colonel
LW	Land Warrior
MALE	Medium Altitude Long Endurance
MAOI	Monoamine Inhibitor
MCI	Mass Casualty Incident
MCS	Maneuver Control System
MEG	Magnetoencephalography
MOS	Military Occupational Specialties
MRAP	Mine Resistant Ambush Protected
MRI	Magnetic Resonance Imaging
MST	Motor Sequence Task
NCAA	National Collegiate Athletics Association

NCO	Network Centric Operations
NCW	Network Centric Warfare
NIH	National Institutes of Health
NIR	Near Infrared Light
NIRS	Near Infrared Spectroscopy
NIRVANA	Nissan Iowa Research Vehicle for Advanced Neuroergonomic Assessment
NOD	Night Observation Device
NON	Non-vulnerable
NT	Neglect Time
NTC	National Training Center
OTV	Outer Tactical Vest
PDA	Personal Digital Assistant
PER3	Period 3
PET	Positron Emission Tomography
PFC	Prefrontal Cortex
ps	picosecond
PTSD	Post Traumatic Stress Disorder
PVT	Psychomotor Vigilance Test
REM	Rapid Eye Movement
RESCHU	Research Environment for Supervisory Control of Heterogeneous Unmanned Vehicles
ROI	Region of Interest
ROTC	Reserve Officers' Training Corps
RT	Response Time
SCID	Structured Diagnostic Interview
SD	Sleep Deprivation
SDSU	San Diego State University
SEALs	Navy Sea Air and Land Forces
SFT	Square Feet
SPECT	Single Photon Emission Computed Tomography
SSRI	Selective Serotonin Reuptake Inhibitor
SWA	Slow Wave Activity
SWS	Slow Wave Sleep
T	Tesla
TCSPC	Time-Correlated Single Photon Counting
TDT	Texture Discrimination Task
TGR	Tactical Ground Reporting
TMS	Transcranial Magnetic Stimulation
TOC	Tactical Operations Center
TRADOC	Training and Doctrine Command
TSD	Total Sleep Deprivation
UAV	Unmanned Air Vehicle
UCOFT	Unit Conduct of Fire Trainer

UCSD	University of California San Diego
UUV	Unmanned Underwater Vehicle
UV	Unmanned Vehicle
VMC	Visuomotor Control Task
VNTR	Variable Number Tandem Repeat
VUL	Vulnerable
WCST	Wilson Card Sort Task
WM	White Matter
WPS	Wi-Fi Positioning System
WRAIR	Walter Reed Army Institute of Research

About the Editors

Dr. Steve Kornguth is the Director of the Center for Strategic and Innovative Technologies and Biological and Chemical Defense, Institute for Advanced Technology at The University of Texas at Austin. He is also Professor in the Department of Pharmacy at The University of Texas at Austin and Professor Emeritus in the Departments of Neurology and Biomolecular Chemistry at The University of Wisconsin at Madison. Additionally he is a member of the Army Science Board.

Dr. Kornguth's research at Wisconsin related to neural development, autoimmune diseases and development of binding agents and platforms for sensors and magnetic resonance image contrast materials. His research efforts at Austin relate to sustaining high-tempo operations performance of soldiers and developing technologies for defense against biological threats.

Dr. Kornguth has also established a team of researchers from UT Austin, Baylor College of Medicine, The U.S. Military Academy, Army Research Laboratory/HRED, and the Iron Horse Brigade (1 BCT) First Brigade Combat Team of the First Cavalry Division, to investigate the physical and cognitive correlates of high-tempo operations activity. This research includes identifying the neurophysiological markers of attentiveness, monitoring brain activity during periods of high and low vigilance, and implementing novel protocols to improve performance in high-tempo environments.

Dr. Rebecca Steinberg is Program Manager and Postdoctoral Fellow in the Center for Strategic and Innovative Technologies (CSIT) and the Institute for Advanced Technology (IAT) at the University of Texas at Austin. She oversees and contributes to a multidisciplinary multi-institutional study examining the effects of 36 hours' total sleep deprivation on cognition and physical performance of U.S. military cadets, soldiers, and University of Texas undergraduates. Rebecca publishes a monthly newsletter detailing recent findings in the field of human performance, which can be found online at http://www.csit.utexas.edu/newsletter/newsletter.lasso. Rebecca received her PhD in Neuroscience from the University of Texas at Austin, 2007, with a focus on neuroendocrinology and toxicology/pharmacology. Her dissertation research examined the effects of environmental pollutant exposure across multiple generations using hormone assays, immunocytochemistry, behavior, and gene expression arrays.

Dr. Michael D. Matthews is currently Professor of Engineering Psychology at the United States Military Academy, where he serves as Director of the Engineering Psychology Program. He is a former Air Force officer with tours of duty at the U.S. Air Force Human Resources Laboratory and as a faculty member at the U.S. Air Force Academy. Dr. Matthews was selected as a Templeton Foundation

Positive Psychology Fellow and much of his research focuses on applying Positive Psychology principles to military contexts. He is on the science advisory board for the Military Child Education Coalition, and served as President of the American Psychological Association's Division of Military Psychology from 2007 to 2008. Collectively, his research interests center on soldier performance in combat and other dangerous contexts.

Notes on Contributors

Mark Beeman, Department of Psychology, Northwestern University.

Colonel Steven Chandler, Deputy Director for Reserve Affairs, Army Capabilities Integration Center.

Mary L. Cummings, PhD, Aeronautics and Astronautics Department—Massachusetts Institute of Technology.

Sam A. Deadwyler, PhD, Department of Physiology and Pharmacology, Wake Forest University School of Medicine.

David F. Dinges, PhD, Professor and Chief, Division of Sleep and Chronobiology, Director, Unit for Experimental Psychiatry, Department of Psychiatry, University of Pennsylvania School of Medicine.

Andrew K. Dunn, PhD, Biomedical Engineering Department, The University of Texas at Austin.

Colonel Karl E. Friedl, Telemedicine & Advanced Technology Research Center, US Army Medical Research and Materiel Command, Fort Detrick, MD 21702–5012.

Richard J. Genik, II, PhD, Wayne State University.

Namni Goel, PhD, Division of Sleep and Chronobiology, Unit for Experimental Psychiatry, Department of Psychiatry, University of Pennsylvania School of Medicine.

Richard Gonzalez, PhD, Department of Psychiatry—University of Michigan.

Robert E. Hampson, PhD, Department of Physiology and Pharmacology, Wake Forest University School of Medicine.

Bradley, D. Hatfield, PhD, Department of Kinesiology, School of Public Health, University of Maryland.

Amy Haufler, PhD, Department of Kinesiology, School of Public Health, University of Maryland.

Colonel Robert R. Ireland, MC, USAF, Office of the Assistant Secretary of Defense (Health Affairs).

Israel Liberzon, MD, Department of Psychiatry—University of Michigan, VA Ann Arbor Health System.

Steve Kornguth, PhD, The Center for Strategic and Innovative Technologies, The University of Texas at Austin.

John Kounios, PhD, Department of Psychology, Drexel University.

W. Todd Maddox, PhD, Department of Psychology, The University of Texas at Austin.

Michael Matthews, PhD, Engineering Psychology Division, the US Military Academy at West Point.

Lieutenant Colonel James L. Merlo, PhD, the U.S. Military Academy at West Point.

Carl E. Nehme, Aeronautics and Astronautics Department, Massachusetts Institute of Technology.

Martin Paulus, MD, The University of California—San Diego; Naval Special Warfare Center.

Eric G. Potterat, PhD, Naval Special Warfare Center; Optibrain Consortium, San Diego, CA.

Matthew Rizzo, MD, Professor of Neurology, Engineering, Public Policy, and Director of the Division of Neuroergonomics, The University of Iowa Carver College of Medicine.

Matthew Rocklage, Department of Psychology, The University of Texas at Austin.

David M. Schnyer, PhD, Department of Psychology, The University of Texas at Austin.

Joan Severson, Digital Artefacts.

Alan N. Simmons, PhD, The University of California—San Diego.

Annette Sobel, MD, MS, Adjunct Professor in Electrical and Computer Engineering and Family and Community Medicine, University of Missouri; Major General (Retired) Arizona National Guard.

Judith L. Swain, MD, University of California, San Diego; Singapore Institute for Clinical Sciences (A*Star)/National University of Singapore; OptiBrain Consortium.

Rebecca M. Steinberg, PhD, The Center for Strategic and Innovative Technologies, The University of Texas at Austin.

Robert Stickgold, PhD, Center for Sleep and Cognition, Beth Israel Deaconess Medical Center and Harvard Medical School, Boston, MA.

Logan T. Trujillo, PhD, Department of Psychology, The University of Texas at Austin.

Karl F. Van Orden, PhD, Naval Health Research Center; OptiBrain Consortium, San Diego, CA.

Acknowledgements

The 2nd Annual "Sustaining Performance Under Stress" Symposium was organized by the Center for Strategic and Innovative Technologies (CSIT) at the University of Texas at Austin through generous funding provided by the Army Research Laboratory, #W911NF-08-2-0015.

The research of Mary L. Cummings and C.E. Nehme was sponsored by the Office of Naval Research, Charles River Analytics, Inc., and Lincoln Laboratory. Special thanks to Dr Birsen Donmez for the statistical support and Yale Song and Dr Jacob Crandall for the test bed support.

The research of Matthew Rocklage, W. Todd Maddox, Logan T. Trujillo, and David M. Schnyer was funded by the Center for Strategic and Innovative Technologies at The University of Texas at Austin with funds provided by ARL-HRED, #W911NF-07-2-0023.

David Dinges and Namni Goel would like to acknowledge support by the NIH NR004281 and CTRC UL1RR024134, by the National Space Biomedical Research Institute through NASA NCC 9-58, and by the Air Force Office of Scientific Research Grant FA9550-05-1-0293.

The work of Robert Stickgold was supported by grant MH48,832 from the National Institutes of Health, USA.

Robert Hampson and Sam Deadwyler would like to thank Ashley Morgan, Kathryn Gill, Joshua Long, Joseph Noto and Santos Ramirez for technical assistance. Drs Linda Porrino, James Daunais and Terrence Stanford collaborated on portions of this project. Their financial support was provided in part by the National Institutes of Health and the Defense Advanced Research Projects Agency. Ampakine CX717 was provided by Cortex pharmaceuticals.

Foreword

Robert E. Foster

As in the past, thousands of young people in the military uniforms of the United States have been asked, today and for the past five years, to dedicate their human performance capabilities to missions and goals set by the United States. This is the clear context for the implications and applications of this book—the current high tempo National Security missions and tasks that have been given to the Military Services of the United States. While human performance is the *sine qua non* of military success, all too often the capabilities of war machines dominate public discussion. It is very appropriate, then, that the conference from which this volume evolved exemplifies the needed national discussion from a multidisciplinary view of the enabling human sciences and a view of the practical challenges to optimal performance in diverse settings and environments.

Human performance has several military contexts that always have to be considered. First is the fact that the US military typically moves small towns (3000+ in population) of young, healthy people to far-flung areas of the globe. These towns are configured as combat brigades, air wings, or battlegroups and are complete with all of the services and support systems found in small municipalities (food/water/sanitation public services, health system, spiritual support, morale and welfare support, etc.). Second, these large populations are organized and trained to execute their missions and tasks and to accommodate the natural variance in human skills and abilities. Thus, a base level of human performance is assumed. Third, the individuals comprising the organizations and the organizations themselves (with standing or standard operational procedures) have been trained and equipped such that some resilience in individual or collective performance can be assumed. These contexts have to be considered in the light of challenges brought on by the realities of the missions and tasks—realities that include extreme environments (heat, cold, altitude and their combinations), extreme physical demands (carry heavy loads), stress (lack of sleep, mental stress, complex cognitive tasks). With this view one can come to the conclusion that a key aspect of human performance has to be cognitive readiness and thus an understanding of the contribution of central nervous system functions to optimal human performance is essential.

Chapters 2–18 in this book give us a broad look at the state of research that can address human performance under the stress of the work and environments in military operations. It is entirely appropriate that the contributions cover methods for assessing performance, cognitive science, stress modalities, and opportunities for performance optimization. After considering this offering, the reader may want to look at other recent compendiums of work related to human performance such as the conference proceedings of the first annual Sustaining Performance Under Stress Symposium (2009), Performance Under Stress (2008), Countermeasures for Battlefield Stressors (2000), Nutrient Composition of Rations for Short-Term, High-Intensity Combat Operations (2006), Opportunities in Neuroscience for Future Army Applications (2009) and the various works that are referenced in the chapters. Finally, I want to acknowledge Steve Kornguth's passion to bring the discipline of neuroscience and its allied physiological sciences to bear on the many issues that face us in optimizing the performance of our military service members who work tirelessly to protect the United States.

References

Committee on Opportunities in Neuroscience for Future Army Applications (2009) "*Opportunities in Neuroscience for Future Army Applications*", Board on Army Science and Technology, National Research Council of the National Academies (Washington, DC: The National Academies Press).

Committee on Optimization of Nutrient Composition of Military Rations for Short-Term, High-Stress Situations (2006) "*Nutrient Composition of Rations for Short-Term, High-Intensity Combat Operations*", Food and Nutrition Board, Institute of Medicine of the National Academies (Washington, DC: The National Academies Press).

Friedl, K., Lieberman, H., Ryan, D.H., and Bray, G.A. (Eds.) (2000) "*Countermeasures for Battlefield Stressors*", 10 (Baton Rouge, Louisiana: Louisiana State University Press).

Hancock, P.A., and Szalma, J.L. (Eds.) (2008) "*Performance Under Stress*", Hampshire, England: Ashgate Publishing Limited.

Kornguth, S.E. and Matthews, M.D. (2009) "Sustaining Soldier High Operations Tempo Performance", Sustaining Performance Under Stress Symposium, *Military Psychology* 21 Supplement 1 (Austin, Texas: Routledge, Taylor & Francis Group).

Chapter 1
Introduction

Steve Kornguth

The chapters for this book were prepared and written at a time when the United States military community is involved in two major confrontations (Iraq and Afghanistan), faced with emerging threats in several regions including Iran and North Korea and with an increased incidence of mental distress or suicide in our young Soldiers and in the student population of our Universities. Concomitantly there has been an explosive growth in the technologies that permit non-invasive examination of brain activation related to neural information processing. This technology that allows visualization of brain activity associated with increased/decreased situation awareness and cognitive capabilities, can be explored as we search for methods to improve Soldier performance and mental health during a period of increasing international stress.

The opportunities provided by technological growth and the health care needs of our young students and soldiers has led to several new studies of interest to our research community. The National Research Council of the US National Science Academy has completed a study describing "Opportunities in the Neurosciences for the Army." The Human Dimension initiative from the Army discusses the perspective of the Soldier and her/his capabilities as a fundamental element of Force Projection. The news media have discussed extensively the mental health issues including suicide that affect student and Soldier wellness. Secretary of the Army Peter Geren has been proactive in calling for and funding a study of suicide through the National Institute of Mental Health that will begin in 2009. The topics addressed by this year's conference are timely and on target for our military forces.

The First Annual Conference on Sustaining Performance Under Stress (Kornguth and Matthews 2009) discussed the need for obtaining a better understanding of Soldier situation awareness under the stress of fatigue, combat and high consequence decision-making (loss of life from friendly fire, errors in judgment and the fog of war). The transition from a heavy armored force to a light agile force and the increasing dependence of our Army on processing increasingly larger amounts of data in very short time intervals highlighted the need to better fuse and present critical information in a manner comprehensible to an operator. It also was deemed essential that some objective measure of operator comprehension be developed. The first Conference is now available as a proceedings in Military Psychology (Kornguth and Matthews 2009).

The major components of this current volume have a focus on neurophysiologic correlates of stress including sensing of the environment under stress, the identification of devices that measure or modulate stress, and lastly, the transition of current research finding to the user. In all sections of this book an effort is made to emphasize the need for objective, nonintrusive, and quantitative measures of an operator's ability to comprehend sensory input. These objective measures can provide a normative database that an operator will use to identify actionable responses appropriate given the fused data (information).

It may be instructive to identify here problems that our Soldier perceives in extended deployment to threat regions (as indicated by Col. James Merlo—see his chapter). We can then try to understand neural mechanisms that may ameliorate the perceived difficulties. The problems identified include the observations that:

1. The Soldier in 2009 operates at the cusp of extreme stress, pressure and sleep deprivation.
2. The connectivity of the soldier in the field (loss of texting, email, Facebook) is markedly different from that existing prior to deployment.
3. Data presentation is not standard. Displays from different manufacturers are not standardized and that causes an operator to switch quickly between operating systems, creating an unnecessary cognitive burden on the warfighter.
4. There is a clear need to identify large numbers of persons in the immediate vicinity who represent differing threats. There are multiple uniformed persons who are friendly and others who are hostile. The same is true of the non-uniformed persons.
5. The rapid changes in threat conditions as experienced during full spectrum warfare operations. In these situations a Soldier may be assisting Iraqis for some period of time and shortly thereafter have a need to manage an IED event and then again to return to reconstruction activities in minutes. There may be a limit to the ability of the brain to shift from peacekeeping to combat and then back to reconstruction. Exceeding that limit could engender cognitive dissonance and dysfunction of complex decision-making.The stressors affecting our deployed Soldiers are diverse. They range from sleep deprivation fatigue, to fear, to loss of limb or loss of a fellow Soldier. Working on the edge of maximal efficiency for long periods of time (12 months) with restricted sleep patterns will affect simple and complex decision-making. Such changes may lead to increased vulnerability or increased resilience in our Soldiers.

The first section of the book correlates individual differences in neural structure and function with behaviors. The potential influence of genetic makeup, training, and endocrine function on performance is examined. David Dinges has utilized vigilance attention tasks to determine whether the vulnerability of individuals to sleep deprivation has stable trait-like properties indicating a possible genetic component. The vigilance attention performance was found to be associated with particular genes having variable-number tandem repeat polymorphisms. The studies reported by Schnyer, Trujillo, and Maddox, also suggest that resilience of an individual to fatigue induced by sleep deprivation is a stable trait. They observed that individual resistance to fatigue from sleep deprivation appears to be a direct function of the diffusion tensor imaging (DTI) pattern (reflective of axonal fiber bundle caliber, degree of myelination and axonal fiber density). Although DTI is a relatively stable trait (the number of axonal processes in the adult and their caliber are not expected to change in short periods of time) the genetic contribution has yet to be determined. These observations imply that there is significant individual variation in the length of time of sleep deprivation required for loss of vigilance to be detected.

Dinges and colleagues observed that for a given subject the total amount of sleep obtained per day was the determinant factor in the psychomotor vigilance attention task performance. When learning and memory consolidation was examined as the critical factor in performance, Robert Stickgold and colleagues observed that slow wave sleep stabilizes recently acquired declarative and procedural memories and REM and stage 2 non-REM sleep enhance such memories. Hence differing sleep measures are required for vigilance attention tasks as compared with memory consolidation. Studies done earlier by Dr Gregory Belenky of the US Army (Belenky et al. 2003, and McLennan et al. 2005) and others revealed that a given individual's performance decays at different rates during sleep deprivation for different specific tasks. The error rate for detection of perimeter penetration increases markedly after three hours time on task even while the ability of a marksman to accurately hit a single target remains stable for over 30 hours of sleep deprivation. However the accuracy of a sharpshooter to differentiate between actual targets and neutral potential targets degrades within 20 hours of sleep deprivation. Therefore individuals vary in their capabilities for sustaining situation awareness and decision-making. Some capabilities may be genetic and would not readily be amenable to training regimes; other capabilities may be improved by training. Also a given individual's capability to detect change and execute proper decision-making will be altered with continued time on task. All the points considered in this paragraph compel us to be aware that "Performance Assessment" is not a unitary factor but depends upon the specific activity being examined.

While the high spatial resolution technology of blood oxygen level-dependent (BOLD) functional magnetic resonance imaging (fMRI) provides a "gold standard" of regional changes in blood oxygen level associated with task performance and stress loading, the utility of this technology in the field is limited. Emerging technologies, including Near Infra Red Spectroscopy (NIRS) offer a potential for lightweight, low energy requiring assessment of brain activation that can be made field deployable. Andrew Dunn discusses the potential utility of NIRS.

In addition to the stable neural anatomical factors that confer resilience or vulnerability to sleep deprivation, there are state changes in regional activation of the brain and in hormonal levels (cortisol) that are associated with complex decision-making in high-risk environments. Marked reduction in the activation of the right pre-frontal cortex is associated with decreased performance on complex decision-making tasks and rapidity of response to emerging threat (Rocklage et al.). Liberzon has presented elegant studies indicating that aversion of an individual to perceived "loss" can be modulated by administration of cortisol and has demonstrated by neuroimaging technologies that the insula region mediated the cortisol effects.

In the majority of studies reported here, the visual system is the dominant sensory input to the human subject studied. The question then arises whether multimodal sensory inputs can sustain operational effectiveness and permit early identification of emergent threats. The chapter written by Lieutenant Colonel James Merlo describes the utility of haptic stimulation (vibratory elements on a belt placed on the abdomen) and demonstrates that multi-modal signaling enables an operator to control which stimulus is perceived first. Merlo and colleagues also demonstrate that such multi-modal signaling is an effective form of performance enhancement in operational conditions. In a related manner it has been shown that combined visual and auditory inputs can markedly improve decision-making in a skilled anesthesiologist (Sanderson et al. 2008, not contributing to this volume) and delay the effects of attention loss.

The studies described above correlate brain activation patterns (from imaging) with behavioral responses in a single specific tasking environment. There is intense interest in the manner by which the brain processes multi-tasking inputs in a real or constructed naturalistic environment. Two of our contributors, Matthew Rizzo and Richard Genik describe their studies of humans in automobile driving tasks. They measured neural responses to emergent threats and distractions during this exercise. In the studies reported, Genik observed an inverse correlation between activation of the right parietal cortex (determined by magnetoencephalography and fMRI) and reaction time to visual cues when a secondary auditory task was added to the environment. Changes in parietal cortical activation were observed in Soldiers and West Point Cadets when they perceive emergent threats (The Center for Strategic and Innovative Technologies-led studies). The Soldiers and Cadets were imaged by fMRI in the alert and in a sleep deprived state as they are visualizing driving through a hostile environment in a humvee. At different times during the video presentation, the subject is told to press a button at the first appearance of potential threat and then another button as the threat is realized. Delay times and accuracy of detection are recorded. Individuals not exposed to combat areas exhibit marked decrease in parietal activation after 24 hours of sleep deprivation. Genik and the team of Schnyer, Maddox, and Trujillo explored multitasking events associated with subject response and observed changes in parietal cortical activation.

A subset of the analyses of performance in naturalistic environments asks the question whether the affect of high performing individuals impacts their functional capability. Martin Paulus and his team investigated by fMRI the activation of the amygdala and insular cortex during an emotional face processing task. If the affect of skilled warriors correlates with sensitivity to emotional face processing, a logical extension of research asks whether specific hormones or pharmaceuticals can modulate or alter social behavior in an operational team. Military education and doctrine fosters the concept that socialization and group identification are critical elements in determining small group operational success. Future experiments validating the observation that hormones can selectively enhance sociability would provide new insights for Soldier training and bonding exercises. Robert Hampson's group has extended the approach to include a non-human primate, the Rhesus macaque. With the aid of technologies including positron emission tomography to investigate metabolic changes in the brain associated with cognitive performance in conditions of sleep deprivation, Hampson observed that select pharmacological agents will sustain performance and alleviate cognitive impairment.

Bradley Hatfield explores how the training of highly skilled marksmen can lead to superior shooting performance. This group observed increased stress arousal and activation and cortico-cortical communication between non-motor and motor regions during the stress of competitive shooting trials. With the aid of frontal electroencephalography (EEG) alpha asymmetry feedback (a model developed by Richard Davidson of the University of Wisconsin) the shooters were able to manage the stress-induced arousal related to competition and thereby improve performance.

One area of emerging interest in the field of neuroscience exploration is the "binding problem". This can be visualized as the sudden perception or insight that permits solving a problem instantly (the eureka moment). An excellent discussion of the binding problem provided by Roskies (1999). This experience differs from the deliberate, methodical problem solving discussed in the majority of presentations included in the book. John Kounios discusses the intuition experience from the vantage of functional neuroanatomy (as observed with EEG and fMRI assessments). Establishing measureable and quantifiable metrics for appropriate "intuitive" behavior may be critical in developing new training paradigms that will permit Soldier survival when the time line between change in threat state and action is exceedingly short (seconds to very few minutes).

Our medical and human performance-oriented presenters from the Uniformed Services, Major General (retired) Annette Sobel, Colonel Karl Friedl, Colonel Robert Ireland, Colonel Steven Chandler, Lieutenant Colonel James Merlo, presented discussions on the application of neuroscience insights and technologies to Warrior survival and effectiveness in the military environment. Major General (retired) Sobel described the operational challenges encountered during mass casualty events including chemical, biological, radiological, nuclear and high explosive incidents. She identified metrics that may be used to evaluate performance. Colonel Chandler reported on the Human Dimension in the Full Spectrum Operations Program of the Army Capability Integration Center (ARCIC), a component of the US Army Training and Doctrine Command (TRADOC). The Human Dimension Program acknowledges the Soldier as the centerpiece of the Army and brings together the cognitive, social and physical needs and capabilities of the individual. Colonel Friedl discussed with insight the need to define the limits of human tolerance so that the combined Soldier, equipment and doctrine can be matched to human capabilities in the Human Performance Enhancement programs. He raised the need to be aware of safety and ethical concerns as we develop new technologies to overcome existing biological limitations. Lieutenant Colonel Merlo discussed the utility of a haptic signaler placed on the abdomen to complement visual and auditory sensory input regarding threat detection. Merlo also identified specific threats faced by the Soldier in "insurgency" or force-on-force operations.

One of the aims of the current volume is to inform our military leadership, interested researchers and students that correlating neural-imaging data with behavioral and cognitive assessments of performance will allow development of improved training protocols and field-ready devices to sustain performance and enhance survival. The real-time assessments available in this coming decade include fitness-for-duty based on physiological measures, technologies for enhancing soldier comprehension of data sets, improved leadership training based on neurophysiological outcomes, and analysis of brain areas directly involved in decision-making. The brain analysis can lead to enhancements of cognitive awareness for extended periods of time during high operations tempo and determine brain structural changes associated with trauma that may impair reinsertion of the Soldier into his/her unit or into society.

These new technologies may also present an invitation to attempt to modify brain processing systems and develop a combatant who has perception and decision-making capabilities that exceed the current potentialities of the human brain. While the development of a super-combatant could involve overt manipulation of neural information processing by implanted electrodes, transcranial magnetic stimulation, neural implants or other "advancements," the potentially adverse unintended consequences of such neural manipulation requires careful ethical evaluation and societal discussion.

One result of the current volume is anticipated to contribute to the development of a "Commander's Dashboard", a graphical computer software interface that will enable leaders to monitor in real-time the physiological and cognitive state of their Soldiers. This interactive technology discussed in the final chapter will permit the Lead Officer to view the status of the unit as a whole or of individual Soldiers via real-time information regarding situation awareness, vigilance, and readiness of the brigade cohort to initiate action. The output reflects EEG, periodic measures such as pre- and post-deployment fMRI imaging of brain activity during a standardized task, and static traits such as MRI structural brain anatomy. The Officer may use this information to change duty assignments, administer nutritional amendments or pharmaceuticals, or order rest or counseling for individual Soldiers and units. This technology can provide baseline and longitudinal data on the status of each Soldier that will improve training, aid in the recognition of super-performing individuals, and identify those individuals whose susceptibility to physical or mental stress merits close observation. Combining these data can enable the development of this technology within a 10-year time frame (consistent with recommendations of the Human Dimension effort of the US Army).

References

Belenky, G., Wesensten, N.J., Thorne, D.R., Thomas, M.L., Sing, H.C., Redmond, D.P., Russo, M.B. and Balkin, T.J. (2003) "Patterns of Performance Degradation and Restoration During Sleep Restriction and Subsequent Recovery: A Sleep Dose-Response Study", *J Sleep Res.*, 12, 1–12.

Kornguth, S. and Matthews, M. (eds) (2007) "Proceedings of the Sustaining Performance Under Stress Symposium", *Military Psychology*, 21-S1, Adelphi, Maryland, 2007: Routledge, Taylor & Francis Group.

McLellan, T.M., Kamimori, G.H., Bell, D.G., Smith, I.F., Johnson, D. and Belenky, G. (2005) "Caffeine Maintains Vigilance and Marksmanship in Simulated Urban Operations with Sleep Deprivation", *Aviat. Space Environ. Med.*, 76, 39–45.

Roskies, A.L. (1999) "The Binding Problem", *Neuron*, 24, 7–9, 111–25.

Sanderson, P.M., Watson, M.O., Russell, W.J., Jenkins, S., Liu, D., Green, N., Llewelyn, K., Cole, P., Shek, V. and Krupenia, S.S. (2008) "Advanced Auditory Displays and Head-Mounted Displays: Advantages and Disadvantages for Monitoring by the Distracted Anesthesiologist", *Anesth. Analg*, 106, 1787–97.

PART 1
Cognition During Real-world Activities

Chapter 2
Promises and Challenges in Translating Neurofunctional Research for Army Applications

Richard J. Genik II

Introduction

Behaviors requiring complex integration of several cognitive and motor tasks are core in today's technological Army. The menagerie of mundane information processing magnifies the malady of multitasking, requiring soldiers to sift a tsunami of data for critical elements among an ocean of obfuscation. Physical stress accumulates from typical and mission-specific sources, such as excess fatigue, sleep deprivation, and hypoxic or thermal extremes. These conditions combine to produce a situation that can degrade soldier performance. Traditional behavioral research will uncover sets of conditions under which performance is affected, and models can be built to attempt to predict soldier performance from recordable and measurable quantities. Neurofunctional research, a relatively new area of inquiry, seeks to directly measure brain activity as the essential ingredient in models to predict behavioral performance.

The laboratories at Wayne State study neural correlates of decision-making under stress, with an emphasis on pushing the edge of neurofunctional technology, while maintaining grounded foundations back to the behavioral application under study. The main tools utilized for this research are functional MRI (fMRI) and Magnetoencephalography (MEG). Utilizing fMRI to measure the neural reaction to block-designed stressors, such as flashing checkerboards or audio tones, is excellent for simple neuroscience research or pre-surgical planning to locate functional areas of an average or individual brain. More complex experiments are needed, however, to produce differential information required for practical application of functional research. At each step in added complexity, technology challenges are presented that incorporate electrical, computer, mechanical, and biomedical engineering. The Emergent Technology Research Division addresses issues of the practical application of advanced, next-generation, and generation-after-next neuroscience tools.

We present here a brief description of the fMRI and MEG technologies, emphasizing those points that provide insight, and those that present limits on applicability. Next is a summary of the promises and challenges, including recommendations on when neuroimaging will provide added value to a research program. We then include an articulation of the multi-year experience of translating in-vehicle behavior to an fMRI and MEG environment and carrying out experiments in divided attention between primary and secondary tasks. We conclude by highlighting recently published results where we've shown that neural asynchrony is associated with slower reaction times in multitasking during driving. In this latest published result, we measured the brain activation of 24 subjects in MEG, ten of which were also scanned separately using fMRI, and found an inverse correlation between activity in the right parietal lobe and performance of the primary task, or addition of a secondary audio and language task.

All of these are foundation studies in translating research in naturalistic decision-making for human-machine interactions to discover the underlying brain mechanisms involved. We find that the technology, as promised, allows us to examine differential neural function in multitasking environments, as well as monitor a subject's reaction to individual events atop full-motion video, while presenting challenges that must be overcome in order to study more complex automotive or military applications.

Measuring Brain and Mind with fMRI and MEG

Noninvasive imaging of human anatomy and physiology was born around the turn of the last century, after Roentgen's discovery of x-rays in 1895. Static x-ray radiographs and real-time fluoroscopes revealed insight into structure and function, respectively, without the need for invasive surgical procedures. Today, the descendents of these apparatus include technologies such as Computed Tomography (CT), where the x-ray source and receivers are rotated around the body to produce 3-D images, and angiography, where an x-ray absorbing liquid, injected into the bloodstream through a catheter, is monitored in 2-D real-time. Injection of radio-tracing chemicals into the body is used in Positron Emission Tomography (PET) and Single Photon Emission Computed Tomography (SPECT). Inherent risks exist in use of the ionizing radiation of these techniques, and other imaging modalities have been developed that rely on high frequency sound reflection (Ultrasound), absorption and retransmission of radio waves (Magnetic Resonance Imaging, or MRI), and optical reflection (Near Infrared Spectroscopy, or NIRS). Recording of electrical activity in the human body using surface probes provides insight into muscle and nervous system function using techniques such as Electroencephalography (EEG), Electrocardiography (EKG), and Electromyography (EMG). Finally, near-surface electrical activity in the brain can be measured with the very sensitive magnetic probes of MEG.

Details of the techniques listed are interesting, but beyond the scope of the current treatise, focused on neurofunctional imaging. What is important to know about modern imaging is that it is no longer formally viewed as demarcated between static and real-time modalities; rather, each technique can detect physiological changes that occur on a given time-scale, known as the temporal resolution for that modality, over a given distance (or area, or volume), known as the spatial resolution. The temporal resolution is related to both the time it takes to complete a scan and the contrast mechanism; for example, a whole-head CT may only take a few seconds to complete, but because it is only sensitive to bone and tissue distributions, or presence of injected contrast agents, its temporal resolution for normal brain function is on the order of days.[1] MEG on the other hand is sensitive to neural current and can detect functional changes on the order of milliseconds, though the spatial resolution is lower than for CT. This interplay between temporal and spatial resolution in neuroimaging is usually cast in terms of "form and function," equivalently "anatomy and physiology," and colloquially "brain and mind." Respectively, these terms refer to insight to be gained, given a resolution point in the spatial and temporal plane. Several example technologies are shown in Figure 2.1.

Functional (f)MRI is a technology that allows monitoring of blood flow in the brain. If one considers MRI to be a magnetic picture of the brain, fMRI is a magnetic movie. When neurons activate, they consume oxygen and the body responds by moving more oxygenated blood to the region of activity to refuel the neurons. Oxygenated blood looks different on a magnetic picture than deoxygenated blood that has refueled local neurons through oxygen metabolism. By snapping several magnetic pictures of the whole brain in rapid succession, our fMRI magnetic movie can be analyzed offline to determine where neural activity occurred in response to an external stimulus such as viewing an image. This is known as Blood Oxygen Level Dependent, or BOLD fMRI. Although other types of functional measurements exist in research, 95 percent of exams utilize the BOLD technique. In general, BOLD fMRI is currently considered synonymous with fMRI.

Utilizing fMRI, researchers can view the brain networks activated during performance of one or several simultaneous tasks, providing an objective measure related to workload that is independent of subjective post-hoc self-evaluations. This technology, coupled with previous work identifying brain networks associated with multitasking, is intended to complement the current behavioral performance evaluation methodology rather than replace it. Indeed, a deep understanding of the behavioral reaction to the stimulus in complex tasks is essential to the analysis of functional images since task-related activity is essentially at the level of noise within the nervous system, and can only be separated by statistical techniques and *a priori* knowledge of knowing when to look.

---

1 Non-normal function, such as ischemic stroke where a vessel is blocked by a clot, can be imaged before and after administration of clot-busting drugs. In these and other clinical pathology cases, functional changes and therefore temporal resolution are measured in tens of minutes.

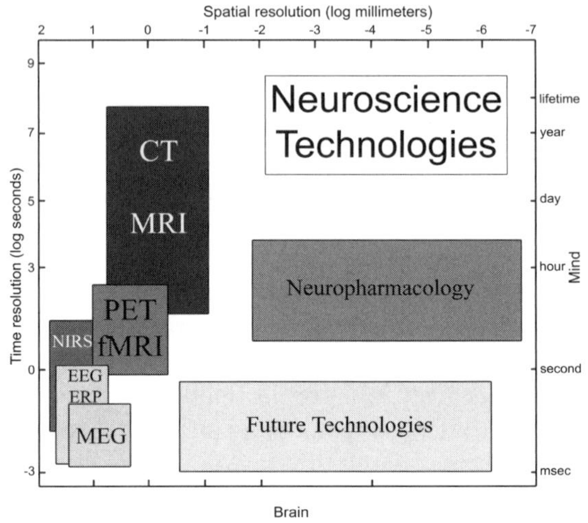

Figure 2.1 Neuroscience technologies provide insight into the brain (anatomy) and mind (function)

The spatial resolution of a given technology defines the largest and smallest brain structures that can be observed, while the temporal resolution defines the elements of mind function to be measured. Most non-invasive technologies are shown, though the list is far from comprehensive. Ongoing research is primarily geared toward improving resolution, though important measurements for the prediction of performance can be at any point in the brain-mind plane. (Adapted from Genik et al. 2005).

Human brains are constantly occupied with autonomic tasks such as breathing and circulatory regulation, and as well as highly cognitive functions such as reading and interpreting technical reports. The human brain does not respond well to a complete reset to a known initial state, like a computer reboot.[2] Therefore, the measured state of any individual brain is a function of its initial state (genetic makeup) and the entirety of internal chemical and external biosensory stimulation integrated from birth to the time the measurement was performed. Measurement of absolute blood flows is therefore highly dependent on the individual. To overcome the extensiveness of variation, one can take a picture of the brain in a baseline state and immediately afterward apply some external stimulus and take a second picture in the activated state. The difference between these images will reveal which parts of the brain activated to confront the stimulus. This is the principle behind functional neuroimaging, and the "differencing view of the world." Such measurements are highly dependent upon

2 The clinical term for nervous system power down preceding a reboot is known as brain death, a state from which there is, so far, no return.

the ability to isolate the subject from uncontrolled stimulation, and additionally establish a baseline brain state in which to insert a stimulus of interest to evoke brain reaction. This makes the creation and understanding of a control condition the foundation of subsequent, more interesting stimulation scenarios. Nearly all activation maps produced in fMRI compare the so-called probe condition to a given control condition, and thus their interpretation is therefore limited. This is one of the major challenges of fMRI in general, and especially for translating the technique to provide something of use for the field. It is important to note that the peak signal from the blood response occurs 4–6 seconds after the neuronal firings, limiting the applicability of BOLD measurements for time-critical paradigms—those requiring interpretation of changes in brain state within around 300 ms. This delayed reaction, colloquially termed the hemodynamic response function *(hrf)*,[3] is another challenge of the technology. The promise of fMRI is unparalleled spatial resolution in whole-head non-invasive functional measurements.

It is worthwhile here to differentiate between experiments that insert a stimulus of interest to evoke a brain reaction, and experiments that insert a stimulus to evoke a brain reaction of interest. This subtle distinction is essentially the primary difference between clinical and research neuroimaging. In clinical fMRI, one uses a stimulus to provoke a robust brain reaction in a known functional area of interest, such as language. This paradigm may ask a subject to think of action verbs to describe a picture. The purpose of such a scan is to examine an individual brain area to see if it is functioning normally. In research fMRI, the brain reaction to a stimulus of interest is the goal, and this makes research fMRI considerably more difficult since the probable areas of activation are only hypothesized *a priori*, and the robustness of strength and locations across subjects can be considered completely unknown before pilot data is acquired. The current experiments in performance are all concerned with this type of research fMRI. The equivalent technology developed for military field use needs to be as robust, or likely more robust than clinical fMRI.

A simple example of an fMRI paradigm includes successive MRI scans while the subject taps his index finger against his thumb for a time interval, the *stimulus interval*, and then remains still for an equal period of time, the *baseline interval*. The stimulus-baseline pairs are repeated several times to obtain a stable average and the difference between the two averages will reveal the neural activity associated with finger tapping—the *neural correlates*. The stimulus design in the previous example is known as boxcar or block design. This design contrasts with an event related design where stimuli are presented or tasks are performed at irregular intervals (Figure 2.2). The averaging of the data is more involved in event-related designs although the result is the same—a mean measurement for each functional state where the difference reveals the neural correlates.

3 The *hrf* refers to the hemodynamic response function, the mathematical relationship between stimulus onset and blood response.

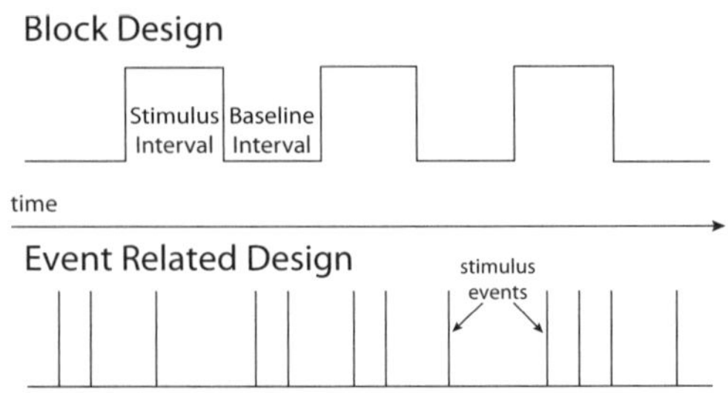

Figure 2.2 Experimental design paradigms for fMRI

In the block design, signals acquired the stimulus condition (convolved with a canonical *hrf*) are statistically averaged and contrasted with a similar average for the baseline conditions. In event-related designs, the delta, or "stick" functions representing event timing are convolved with an *hrf*. In a simple experiment, events of two types would be presented and their relative statistical presence in the observed signals generates the activation maps.

MEG experiments can be conducted in very much the same manner as fMRI. The main promise of MEG is the high temporal resolution, though sufficient spatial resolution facilitates fusing information from the two modalities. Hence, another promise is a multimodal temporal resolution of milliseconds combined with the spatial resolution of millimeters or less. Of course, detailed methods of the combination are a major challenge and current research does the analyses in separate steps, using a traditional MEG approach of frequency power analysis or locking an average signal to the onset of an event cue to search for an event-related localized activity peak, in a very similar fashion to Event-Related Potentials (ERPs) from EEG.

Promises and Challenges

For continuous activity, such as attending to multiple sensory inputs in a Humvee cockpit, block design experiments provide a robust design and good signal to noise ratio. These paradigms also translate well between fMRI and traditional MEG/EEG power frequency analysis as the *hrf* has been shown to relate to localized coherent gamma oscillations (Logothetis et al. 2001, and Muthukumaraswamy and Singh 2008). These promises are tempered by the fact that this type of analysis only works for continuous activity, and it is hypothesized that the bulk of interesting paradigms that parallel naturalistic military environments will show performance degradation related to transient signals, such as failure to register a change in the environment or perceive a warning signal.

To address the analysis of neural activity arising from transient signals, event related paradigms must be considered. These are challenging because of lower signal to noise ratio, required hardware synchronization, and difficulty in establishing a well-understood control condition. These paradigms are additionally sensitive to *hrf* models, and normal inter- and intra-brain differences. On the positive side, using the event locked analysis techniques of fMRI and MEG time series, locating the transient signals is straightforward once the challenges are addressed.

Both modalities, fMRI and MEG, are expensive, with several thousand dollars required per subject for scanner time and data processing costs for a typical study. Additionally, the subject's head must remain still, ideally with less than 0.4mm of positional deviance. Moreover, shielded rooms and special magnetic conditions are required for data collection, conditions that are not available in field environments. Finally, in order to observe overload conditions, 90 percent cognitive utilization is required, which limits the scan time to around seven minutes and 4–5 repetitions, or about 25–35 minutes of total functional scan time. Exceeding these limits introduces cognitive fatigue as a major confounding effect.

With all of these challenges, it is proper to ask "where is there a value-add to utilize neuroimaging over traditional behavioral experiments?" There are three major cases where it is useful in non-clinical situations to utilize brain imaging:

1. In the case where one does not understand or cannot model the behavior and a "ground truth" objective measure of brain network activity or population differences is needed. This has direct application to correlate biomarkers with brain signals (see chapters in this book by Friedl and by Steinberg, Matthews, and Kornguth for detailed examples).
2. In the case where one expects subtle differences in behavior that cannot be accurately measured before it is too late. Information overload, for example, is a condition that is much more useful to predict than observe.
3. In the case where one suspects that more than one neural effect is causing the observed behavior. In this case one would hypothesize neural activation in different networks for different effects, and the high resolution of fMRI should be able to separate them.

Other cases are useful for academic understanding of effects, but in an application-oriented environment, neuroimaging is a more expensive way to record the same reaction time information available in a far less restrictive and typically far less expensive behavioral setting.

Driving and Workload

Driving is an integrated multitask behavior engaging several processes requiring the performance of different but interrelated skills that rely on interconnected visual, motor, and cognitive brain systems, and the associative centers and networks that combine them, similar to complex activities that are performed in military environments. Driving is also a common activity, regularly performed under various stress conditions and imposing differential workload on the subject. There is extensive behavioral and burgeoning brain research in the performance of this task under different conditions, for example including visual distraction and alcohol-induced behavior modification (Allen et al. 2009, Carvalho et al. 2006, and Graydon et al. 2005). Research in driving can be used as a model for aspects of soldier performance beyond the obvious convoy and transport tasks. Driving itself is an opportunity for leveraging research funding with civilian agencies in basic and applied areas and it has the added advantage of being easily understood by civilian leaders as having real-world importance.

Translating between the two research agendas, civilian and military, is made easier by the commonality of the cognitive task: both include a multitasking environment where the performance of over-learned skills is recorded during periods of sustained vigilance. Situational awareness is also common, where the threats are other vehicles or obstacles in the civilian case, and the military case includes those threats, and adds quite a few more such as enemy detection and ordnance avoidance. The populations appropriate to test in the two environments are quite different, though, especially in size, with the civilian driving population on the order of 100 million, and the military population that may be performing any specific multitasking behavior between a few dozen up to a maximum on the order of 100 thousand. It is still a matter of research whether civilian behavior can model military behavior; that is, whether the results of behavioral and neuroimaging research using civilian, typically university community subjects is predictive of results that would be obtained if soldiers were studied directly.

The major challenge in translating neuroscience research to real-world use is validating that the scenario used in the fMRI or MEG environment is predictive of real-world performance. This is an important step that is typically skipped in the more sensationalized studies, such as Just et al. (2008) where subjects in the baseline driving task averaged "errors" at a rate of eight incidents for every few minutes of simulated driving. An error was defined as running the vehicle into a wall, and this clearly is not even remotely related to actual driver performance; Just et al. was obviously measuring the interference cause by *dual novel task* performance, rather than the effect of audio and language distraction on performance of an over-learned task. A more reasonable approach is to perform a separate behavioral development study to determine which simple tasks in a driving environment are able to predict on-road performance. Then it is required that the development of any neuroimaging scenario include a task that mimics the developed behavioral task that predicts on-road performance. The challenge with this type of bootstrapping is that it requires several years of research to properly accomplish any specific scenario.

The Transportation Imaging Laboratory, a unit of the Emergent Technology Research Division of the Wayne State University School of Medicine, was established in 2002 by an industry grant with a charge to develop an in-magnet scenario for simulating on-road driver performance. Initial work consisted of translating the detect-and-respond task of what is known as the static load test.

The detect and respond task in the static load environment, primarily developed to simulate visual detection of external brake lights and application of braking to the subject's vehicle, but was shown to emulate well the performance on secondary tasks when even speed and lane deviations were taken into account (Angell et al. 2002). The matching of static lab and the developed in-magnet scenario behavioral reaction time mean and variance show that similar performance on the primary task is obtained between the two environments (Young et al. 2005). The resultant brain activation are therefore considered predictive of what is actually involved in

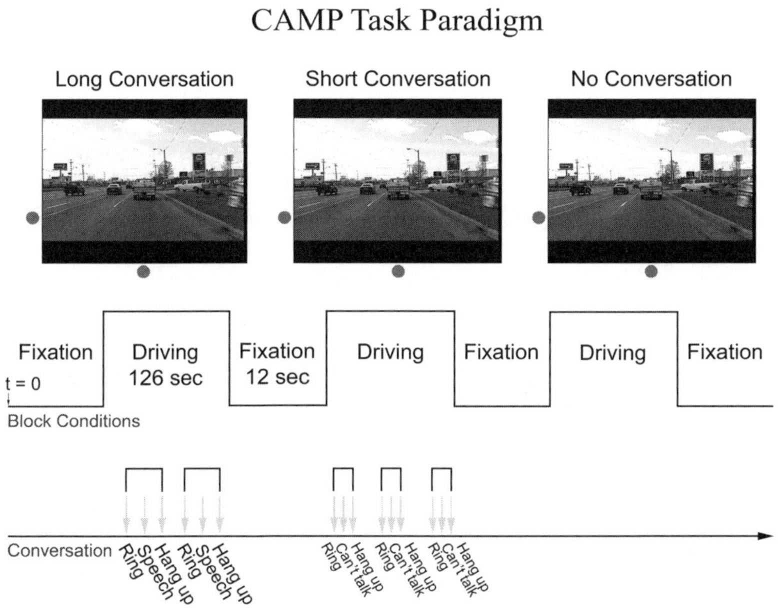

Figure 2.3 Task for the CAMP audio distraction study

Three conversation conditions were studied in each behavioral, MRI, and MEG environments. All conversation conditions contained a background video that ran throughout the blocked driving condition. Fixation blocks were a black screen with a centered fixation cross. The primary detect-and-respond task was the appearance of either a center or peripheral light and followed by the subject depressing a foot switch (simulated brake pedal). For either conversation condition, a ring would sound to which the subject must press a button to answer the call. A long or short conversation ensued and the subject presses the button again to end the call. In the short conversation condition, the subject covertly speaks that they are driving, can't talk now, and they then end the call. In the long conversation condition the subject covertly answers recorded questions such as "What is your birthday?" or "What street do you live on?" Covert speech is used in the MRI and MEG to limit motion artifact. A systematic behavioral study showed reaction times were equivalent in convert and overt speech for this scenario. (Description and figure adapted from Bowyer et al. 2009, and Hsieh et al. 2009).

visual detect and response in real world driving (Graydon et al. 2005, and Young et al. 2005), and that this in-magnet scenario is appropriate for use as a primary task to evaluate the effects of secondary tasks.

Subsequent support from a consortium of manufacturers through the Crash Avoidance Metrics Partnership (CAMP) in 2004–06 addressed the issue of audio and language distraction to the detect-and-respond task in the simulated driving environment. In this study, the primary task of driving and light event detection was modulated by simulated cell phone conversation, as described in Figure 2.3. Complete results of this study are available in the detailed publications (Bowyer et al. 2009, Hsieh et al. 2009). Shown in Figure 2.4 are the MEG imaging results for comparing the simulated driving alone with simulated driving plus long or short conversation. The modulation in the right superior parietal region is expected, as this is where motor and visual integration occurs, though this is the first study to show such modulation in a real-life verified scenario.

The neural correlates underlying multitasks in the CAMP paradigms were coherent between static lab, fMRI and MEG results. We observed activity in the fronto-parietal and visuo-auditory-motor pathways that are known to be involved in visuo-sensory multimodal processing and integration, as well as decision-

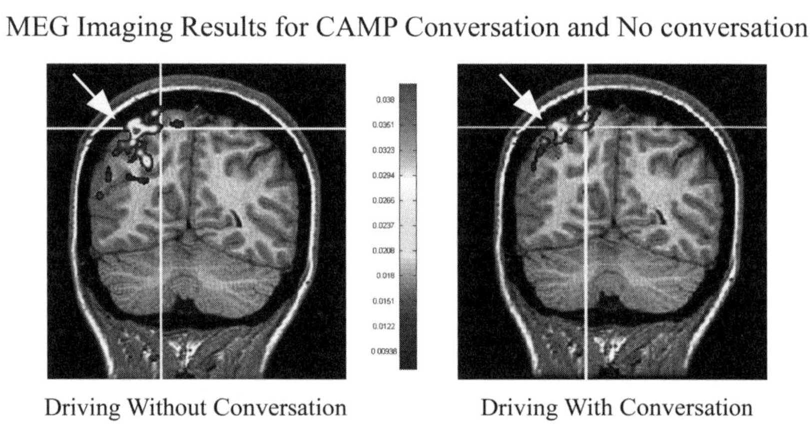

Figure 2.4 Right parietal activation difference in conditions where the subject is attending to the primary driving task (left), and the primary driving task plus the either long or short conversation (right)

These images show strong correlations at 200 ms after the light appears in the driving alone condition, while this correlation is greatly reduced (less areas of high activation) when conversation is added. This results is consistent with observation using the same paradigm in fMRI (Hsieh et al. 2009). The no conversation condition, selected for responses with the longest reaction times, mimics driving with conversation (Bowyer et al. 2009). Note that in the MEG convention, the right side of the brain appears on the left side of the image.

making and attention modulation under a multiple task condition. Furthermore, the use of fMRI together with MEG in investigating simulated driving tasks revealed the neural pathways responsible for cognitive processing in driving-like scenarios.

Our key new finding is that the right superior parietal association region is likely the site of effect for increased cognitive demands while integrating conversation into a driving-like behavior on event reaction time performance. This will be the primary area to study for stress modulation by sleep deprivation, and so on.

To explain our results, we propose a top-down control model where frontal regions influence the synchrony of superior parietal lobe and extrastriate visual cortex that in turn modulates reaction time. This is accomplished by either:

1. Damping brain activation in specific regions during specific time windows;
2. Reducing facilitation from attention inputs into those areas; or
3. Increasing temporal variability of the neural response to visual events (mean reaction time decreases in amplitude for increased variance).

More studies are required to disambiguate between these possibilities.

Conclusions

It has been shown that fMRI and MEG are promising technologies to assist in building and understanding computational models of cognitive processes. These tools are just starting to be applied to military paradigms and many technological challenges still exist; however we have shown that full-motion video and overlaid information and shapes, typical of a military augmented reality display, are feasible and robust. These technologies are not going to be made available in the field in the foreseeable future so lab results will remain an essential ingredient to study the brain regions and networks active during any relevant task. Translation of naturalistic paradigms will not be exact, however, all that is needed is a "critical mass" of behavioral cues and verification that the task performance (including reaction time and variance, for example), in the imaging laboratory match that one observes in more naturalistic environments. The current direction of this research is to use neuroimaging lab results to build models to correlate other biomarkers that can be deployed in the field, such as EEG, eye-tracking, or combined physiological metrics. Finally, it is very desirable to have a behavioral program connection to any neuroimaging studies if the goal is to integrate end-user applications with the results—that is, all translational research in this area must be interdisciplinary, where the end users collaborate with the neuroscientists.

References

Allen, A.J., Meda, S.A., Skudlarski, P., Calhoun, V.D., Astur, R., Ruopp, K.C. and Pearlson, G.D. (2009) "Effects of Alcohol on Performance on a Distraction Task During Simulated Driving", *Alcohol Clin. Exp Res*, 33, 617–25.

Angell, L., Young, R., Hankey, J. and Dingus, T., (eds) (2002) "An Evaluation of Alternative Methods for Assessing Driver Workload in the Early Development of in-Vehicle Information Systems", SAE Government/Industry Meeting, Washington, DC, May 13–15, 2002: Society of Automotive Engineers.

Bowyer, S.M., Hsieh, L., Moran, J.E., Young, R.A., Manoharan, A., Liao, C.C., Malladi, K., Yu, Y.J., Chiang, Y.R. and Tepley, N. (2009) "Conversation Effects on Neural Mechanisms Underlying Reaction Time to Visual Events While Viewing a Driving Scene Using MEG", *Brain Res*, 1251, 151–61.

Carvalho, K.N., Pearlson, G.D., Astur, R.S. and Calhoun, V.D. (2006) "Simulated Driving and Brain Imaging: Combining Behavior, Brain Activity, and Virtual Reality", *CNS Spectr*, 11, 52–62.

Genik, R.J., 2nd, Green, C.C., Graydon, F.X. and Armstrong, R.E. (2005) "Cognitive Avionics and Watching Spaceflight Crews Think: Generation-after-Next Research Tools in Functional Neuroimaging", *Aviat Space Environ Med*, 76, B208–212.

Graydon, F.X., Friston, K.J., Thomas, C.G., Brooks, V.B. and Menon, R.S. (2005) "Learning-Related fMRI Activation Associated with a Rotational Visuo-Motor Transformation", *Brain Res Cogn Brain Res*, 22, 373–83.

Hsieh, L., Young, R.A., Bowyer, S.M., Moran, J.E., Genik, R.J., 2nd, Green, C.C., Chiang, Y.R., Yu, Y.J., Liao, C.C. and Seaman, S. (2009) "Conversation Effects on Neural Mechanisms Underlying Reaction Time to Visual Events While Viewing a Driving Scene: fMRI Analysis and Asynchrony Model", *Brain Res*, 1251, 162–75.

Just, M.A., Keller, T.A. and Cynkar, J. (2008) "A Decrease in Brain Activation Associated with Driving When Listening to Someone Speak", *Brain Res*, 1205, 70–80.

Logothetis, N.K., Pauls, J., Augath, M., Trinath, T. and Oeltermann, A. (2001) "Neurophysiological Investigation of the Basis of the fMRI Signal", *Nature*, 412, 150–57.

Muthukumaraswamy, S.D. and Singh, K.D. (2008) "Functional Decoupling of BOLD and Gamma-Band Amplitudes in Human Primary Visual Cortex", *Hum Brain Mapp*, in press.

Young, R., Hsieh, L., Graydon, F., Genik II, R., Green, C. and Bowyer, S. (2005) "Mind-on-the-Drive: Real-Time Functional Neuroimaging of Cognitive Brain Mechanisms Underlying Driver Performance and Distraction", Proceedings of the Society of Automotive Engineering.

Chapter 3
Modeling the Impact of Workload in Network Centric Supervisory Control Settings

Mary L. Cummings

C. E. Nehme

Introduction

Network Centric Operations (NCO), also known as Network Centric Warfare (NCW), is a concept of operations envisioned to increase combat power by electronically linking or networking relevant entities across a battlefield. As defined by the Department of Defense (DoD), key components of NCO include information sharing and collaboration which will theoretically promote shared situational awareness and overall mission success (Department of Defense 2001). What this means practically is that personnel will have access to exponential amounts of information as compared to today's forces, and thus the information intake for the average NCO operator will be higher than ever before in command and control settings. As a result, mental workload will likely increase, and a large body of human factors literature has demonstrated that as mental workload increases, performance can be negatively affected.

Given that NCO will likely bring about higher mental workload due to the large volume of incoming information, it is critical that systems designers develop predictive models of both human and system performance such that they can determine, given a particular concept of operations, or a proposed technology, what the impact will be on operator workload, particularly in terms of reaching critical performance thresholds. Towards this end, this paper will introduce a new discrete-event simulation approach to human performance modeling that includes a quantitative relationship between workload and performance.

Background

To achieve the vision of NCO, the DOD will increasingly need to rely upon the concept of supervisory control, which is when an operator intermittently interacts with a computer that closes an autonomous control loop (Sheridan 1992). In particular, there is currently a significant thrust towards including unmanned vehicles in military operations, and human interaction with such vehicles, either on the ground or in the air, is inherently a supervisory control problem. Moreover, given increased autonomy of unmanned vehicles (UVs), a human operator's role is shifting from controlling one vehicle in a team of operators to supervising multiple vehicles individually (Department of Defense 2007). However, a primary limiting factor in this one operator—many UV vision, is operator workload. Indeed, this limitation on the control of multiple UVs extends to any supervisory control tasks requiring divided attention across multiple tasks such as air traffic control and even supervisors multitasking in a command center such as an air operation center.

Mental workload results from the demands a task imposes on the operator's limited resources; it is fundamentally determined by the relationship between resource supply and task demand (Senders 1964). While there are a number of different ways to measure workload (Pina et al. 2008, Wickens and Hollands 2000, Wierwille and Eggemeier 1993), given the temporal nature of supervisory control systems, particular to those in command and control settings, we propose that the concept of utilization is an effective proxy for measuring mental workload. Utilization is a term found in systems engineering settings and refers to the "percent busy time" of an operator; for example, given a time period, what percentage of time was that person busy. In supervisory control settings, this is generally meant as the time an operator is directed by external events to complete a task—that is, replanning the path of an unmanned aerial vehicle (UAV) because of an emergent target. For instance, given an operator shift of eight hours, the utilization would be that amount of time the operator was engaged in directed tasks divided by 480 minutes. What is not included in this measurement is the time spent monitoring the system, just watching the displays and/or waiting for something to happen.

The concept of utilization as a mental workload measure has been used in numerous studies examining supervisory controller performance (Schmidt 1978, Cummings and Guerlain 2007, and Rouse 1983), and these studies generally support that when tasked beyond 70 percent utilization, operators' performances decline. While arguably this is not a perfect measure of mental workload, one strength of such a measure is its ratio scale, which allows it to be used in quantitative models. Given the previous research showing that supervisory control performance drops when utilization is greater than 70 percent, we investigated whether using such a numerical relationship could be used to not just describe observed human behavior, but also be used to predict it.

In order to investigate this possible relationship between workload (as measured by utilization) and performance, we elected to use a discrete event simulation (DES) model. A DES model can represent a system as it evolves over time by representation of events, that is, instantaneous occurrences that may change the state of the system (Law and Kelton 2000) and is particularly suited to model supervisory control systems due to their time-critical, event-driven nature which are characteristic of NCO environments. Discrete event simulations are based on the queuing theory, which models the human as a single server serially attending the arrival of events (Senders 1964, Sheridan 1969, Carbonell et al. 1968). These models can also be extended to represent operator parallel processing through the introduction of multiple servers (Liu 1997 2006). The next section presents an overview of the DES model and the inclusion of the utilization-performance relationship.

The Model

In our model, the NCO human operator is responsible for multiple tasks, specifically the management of multiple unmanned vehicles. This operator is modeled as a server in a queuing system with discrete events representing both endogenous and exogenous situations an operator must address. Endogenous events, which are vehicle-generated or operator-induced, are events created internally within the supervisory control system, such as when an operator elects to replan an existing UV path in order to reach a goal in a shorter time. Events that result from unexpected external environmental conditions that create the need for operator interaction are defined as exogenous events, such as emergent threat areas that require replanning vehicle trajectories. All supervisory control systems have such events arriving into the system, often of differing priorities, which will subsequently be addressed.

The design variables that serve as inputs to the model in Figure 3.1 are composed of variables related to the vehicle team (number of vehicles and the vehicle types, as expressed through neglect time), the human operator (interaction times, operator attention allocation strategies, and the operator situational awareness), and a model of environment unpredictability. These are discussed later in further detail.

Vehicle Team Input Variables

In our model in Figure 3.1, we include the number and type of vehicles in the system. While the number of vehicles directly affects the number of events arriving to the operator's queue, we also need to represent the varying types of vehicles, since in the future NCO vision, one person can control multiple types of UVs.

We do this through the concept of neglect time (NT). NTs represent the time a vehicle can operate without human intervention, and different NTs represent different vehicle types.

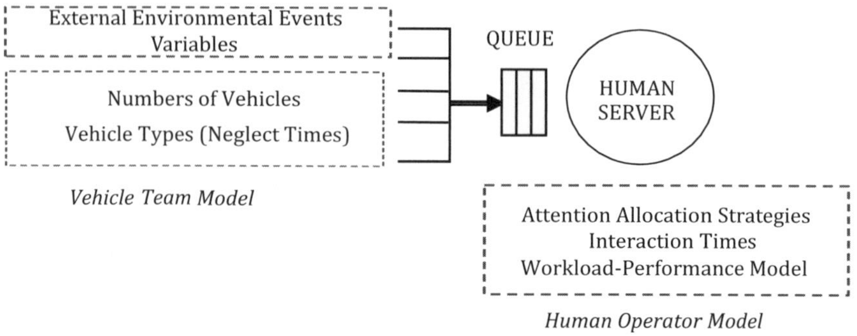

Figure 3.1 A high level representation of the discrete event simulation model

Human Operator Input Variables

The length of time it takes the operator to service an event, also known as interaction time, is captured through a distribution of event service times; thus, the service process can be described by a probabilistic distribution representing the interval from the time the operator decides to service a particular event until the operator is done servicing. We model the operator's strategies for handling priorities, known as attention allocation (Crandall and Cummings 2007a), through the queuing policy, for example, which task waiting in a queue the operator elects to service. Examples include the first-in-first-out (FIFO) queuing scheme as well as the highest attribute first (HAF) strategy (Pinedo 2002). The HAF strategy assumes that high priority events are serviced first, a typical strategy employed in high-risk command and control settings. We also model the operator's management strategy, which is the willingness of the operator to allow the vehicles to operate on their own without intervention. This variable can be seen as a trust proxy since those operators who do not trust the system will intervene more frequently than those who do.

The Workload-Performance Model

As illustrated in a previous study examining single operator control of multiple unmanned vehicles (Cummings and Mitchell 2008), successful models of human performance in such systems should contain some representation of the inherent delays that humans will introduce into these complex systems.

These delays occur through queues that build for multiple tasks since humans cannot instantaneously make decisions, or through delays due to a loss of situation awareness because operators do not realize the system needs their attention. While the delays due to increasing queue size are accounted for in a typical discrete even simulation model, the impact of the loss of situation awareness on a system is far more difficult to both measure and predict. As seen in Figure 3.1, we propose that we can account for the delays in the system due to this loss of operator situation awareness through a workload-performance model.

This workload-performance model is inspired by the Yerkes-Dodson inverted-U relationship (often called a law, which has been a point of contention (Hancock and Ganey 2003), notionally illustrated in Figure 3.2. For our purposes, we will use the previously discussed measure of utilization as our proxy measure for workload. While the original Yerkes-Dodson research focused on stimulus strength and learning (Yerkes and Dodson 1908), a similar effect on the arousal level and performance was found in later work (Hebb 1955). In general, this relationship states that people work best under moderate levels of workload, and that high or low levels of workload will result in degraded operator performance.

The Yerkes-Dodson relationship has been labeled as overly simplistic and more of an appeal to common sense than to scientific rigor (Yerkes and Dodson 1908), which is no doubt true to some degree. However, previous research has shown direct evidence of such a relationship in supervisory control performance using utilization as a workload measure (Cummings and Guerlain 2007). Moreover, when attempting to build models of any system's performance, human or otherwise, Occam's Razor prevails, and if a relatively simple relationship can provide robust predictive results across a number of independent data sets when embodied in a human-system model, then this relationship should not be dismissed, but rather more deeply investigated.

Operator utilization, our measure of arousal, is hypothesized to ultimately affect performance, such that it is degraded at both high and low ends of the utilization curve (Figure 3.2). When operators are under high levels of utilization, it is expected that they can be too busy to accumulate the information that is required to perform at optimal levels. When operators are under-utilized, we propose that due to a low level of arousal and complacency, they can overlook information from the environment or engage in a non-related task (such as reading the paper),

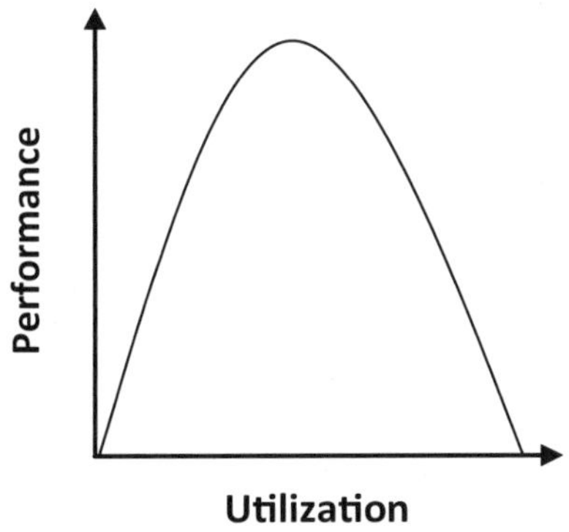

Figure 3.2 The notional workload-performance relationship

which also leads to degraded performance. There is a possible connection between situation awareness and this workload-performance curve, which has been discussed in detail elsewhere (Nehme, Crandall, and Cummings 2008, Donmez et al. *in review*).

Experimental Validation for the Workload-Performance Model

In order to validate the hypothesis that the previously described workload-performance model was necessary in the discrete event simulation (DES) model illustrated previously in Figure 3.1, experimental data from a multiple unmanned vehicle simulation was analyzed. Given the previous debate as to the validity of the inverted-U Yerkes-Dodson relationship, it was critical to test to see whether the inclusion of the proposed parabolic relationship data (Figure 3.2) improved the DES model's ability to replicate observed human experimental data.

Experimental Apparatus and Participants

In this experimental test bed, the operator's mission was to supervise multiple unmanned ground vehicles via a supervisory control interface in order to remove as many objects as possible from a maze in an eight-minute time period. The objects were randomly spread through the maze, the shape of which was initially unknown. However, as each UV moved about the maze, it created a map, which it shared with the subject and the other UVs in the team. There were 22 possible objects to collect during a session. Vehicles contained some level of automatic path planning as well as well as automated goal designation. However, these algorithms were sometimes sub-optimal, which could cause the vehicles to become stuck, thus requiring operator intervention. Typical of heuristic algorithms, they also were sometimes inefficient in goal tasking, which could be significantly improved by human intervention. These kinds of sub-optimal automation behaviors are very common in systems with embedded autonomous planning, as these algorithms are inherently brittle and unable to respond to complex system events (Smith, McCoy, and Layton 1997).

The human-UV supervisory control interface was the two-screen display shown in Figure 3.3. On the left screen, the map of the maze was displayed, along with the positions of the UVs and objects with known locations. This screen also indicated to operators which vehicle was in a particular state and which needed intervention. It also included a time line that indicated mission temporal progress. In order to remove an object from the maze, operators were required to first complete a visual task, which is representative of real world intelligence, surveillance, and reconnaissance missions. To simulate this visual task, users were asked, using the right screen in Figure 3.3, to identify a city on a map of the mainland United States using *Google Earth*-style software.

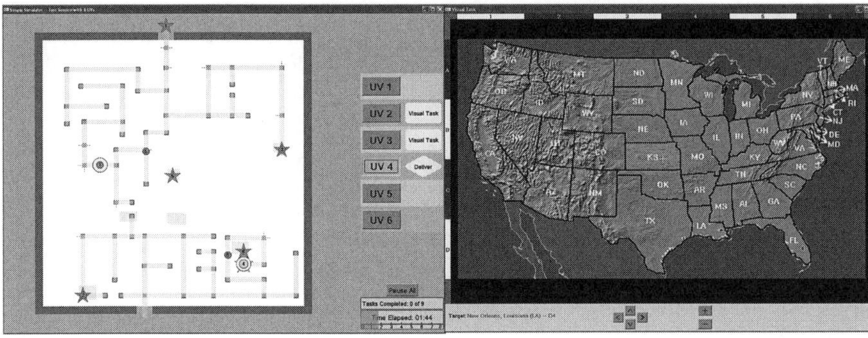

Figure 3.3 Two-screen interface for single operator control of multiple UVs

Sixteen participants between the ages of 18 and 45 participated in this study. The experimental results reported here represent a subset of experimental results from larger studies investigating predictive modeling techniques for supervisory control systems (Crandall and Cummings 2007 a,b,c). Participants completed three comprehensive practice sessions, followed by four test sessions with 2, 4, 6, and 8-vehicle team sizes (randomized and counterbalanced to control for any learning effect. Participants were paid $10 per hour; the highest scorer also received a $100 gift certificate.

Results

In order to compare the DES model results to the experimental results generated from the experiment, probabilistic distributions were obtained for the arrival and service rates as experienced by the participants in the study. These distributions included the service times of the visual tasks as well as the replanning of vehicles between different objects. The inter-arrival time distributions representing those times between autonomous replanning events were also included. In terms of performance, the DES model awarded a point for the servicing of every vehicle-generated event, such as replanning a vehicle when it became stuck. Using the distributions of arrival rates and service times generated from the data in the experiment, 10,000 trials were conducted with the DES, in order to compare the results with the human-in-the-loop experiment. The workload-performance model was a parabolic curve that penalized operators by essentially adding delay time to their interactions, based on the utilization measure. This, in turn, slowed the rate at which participants acquired points. Further details about both the simulation test bed and the model can be found in Nehme (2009).

The observed system performance and operator utilization from the user study are compared with the model's estimates in Figure 3.4. Two dependent variables were measured, the performance, a combination of the objects collected and the correction of erroneous automation behavior, and utilization, as defined previously, the time operators were engaged in tasks divided by the total experimental session time. In Figure 3.4, the results of the human-in-the-loop (HITL) study can be seen in terms of performance and utilization, as well as the results of the model if the workload-performance curve is or is not included.

Without the workload-performance curve, the model tended to over-predict performance for the 4, 6 and 8-vehicle conditions and under-predict for the 2-vehicle condition. Including the workload-performance curve in the DES model led to decreased performance scores (and thus improved predictions), except in the 2-vehicle condition where the model was already slightly under-predicting performance. However, the model's results for system performance when including the workload-performance curve score were all within the 95 percent confidence intervals, except for the 2-vehicle case. The model without the workload-performance curve was well outside the 95 percent confidence interval in all cases.

The inclusion of the workload-performance curve in the model also had an overall positive effect in terms of utilization predictions, as seen in Figure 3.4-b. Without this curve, the DES model tended to over-predict operator utilization for the 4, 6 and 8-vehicle conditions and under-predict for the 2-vehicle condition. When the operator is processing too many tasks, utilization is high which can lead to low performance (as computed by the model). Also, as suggested by Yerkes-Dodson, and confirmed in our empirical results, low utilization translates into fewer correct tasks processed by the operator, thus influencing performance as seen in Figure 3.4-a.

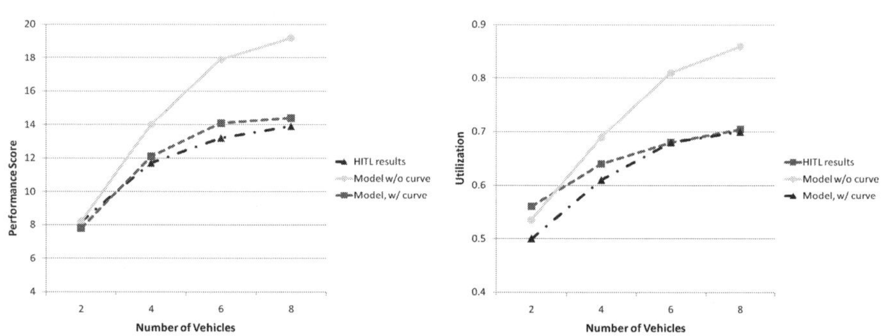

a. Performance Score

b. Utilization

Figure 3.4 Human-in-the-loop (HITL) results versus model predictions

An additional observation about the accuracy of the model can be made from these results, in that the model is more accurate for larger teams than for smaller teams. This trend appears to be caused, at least to some degree, by overly high penalties associated with low utilization in the workload-performance curve in the model. This model parameter leads to the under estimation of both operator utilization and system performance in this condition. This suggests that while the concept of high or low workload contributes to lower operator, and thus system, performance, the symmetric relationship as depicted previously in Figure 3.2 inspired by Yerkes-Dodson, is not quite accurate. This observation was seen again in a second related experiment, detailed in the next section.

Additional Experimental Validation for the Workload-Performance Relationship

Given the results in the previous section, additional experimental data was gathered to assess the nature of the workload-performance curve relationship as depicted in Figure 3.2. The previous section confirmed that including such a relationship in a DES model significantly improved the predictive ability of this model for actual human performance. However, there was an indication that the symmetrical shape of the curve was not accurate, causing the DES model to under-predict utilization and performance. Using data gathered from a completely independent experiment from the one previously discussed, we investigated whether the model, including the workload-performance curve, would also accurately replicate observed HITL performance. In addition, given the previous results, we also wanted investigate the observed shape of the performance and utilization curves. We could not do this in the previous experiment because the experiment was not explicitly designed to investigate this curve; rather, we hypothesized post-hoc that such a relationship was needed after seeing the results.

Experimental Test-bed

The Research Environment for Supervisory Control of Heterogeneous Unmanned Vehicles (RESCHU) simulation test bed allows operators to control a team of UVs composed of unmanned air and underwater vehicles (UAVs and UUVs). All vehicles engage in surveillance tasks, with the ultimate mission of locating specific objects of interest in urban coastal and inland settings. There is a single UUV type, and two UAV types. The first is a high altitude long endurance (HALE) UAV that provides high-level sensor coverage (akin to a Global Hawk UAV) for target identification, while the other, a medium altitude long endurance (MALE) UAV, provides more low-level target surveillance and video gathering (similar to a Predator UAV). Thus, an operator can control up to three different vehicle types. Basic flying and navigation tasks for the different vehicles is highly automated with embedded autonomy to allow operators to concentrate on payload tasking (Cummings et al. 2007).

The RESCHU interface consists of five major sections (Figure 3.5). The map displays the locations of vehicles (U-shaped icons), threat areas (circles), and areas of interests (AOIs, indicated by diamonds). Vehicle control is carried out on the map, such as changing vehicle paths, adding a waypoint (a destination along the path), or assigning an AOI to a vehicle. The main events in the mission (for example, vehicles arriving to goals, or automatic assignment to new targets) are displayed in the message box, along with a timestamp. When the vehicles reach an AOI, a simulated video feed is displayed in the camera window, and the operator then visually identifies a target in this simulated video feed for just MALE UAVs and UUVs. Example targets and objects of interest include cars, swimming pools, helipads, etc. When vehicles complete their assigned visual tasking, an automated-path planner automatically assigns the HALE UAV to an AOI that needs intelligence, and the MALE UAVs and UUVs to AOIs to pre-determined targets. As in the previous experiment, the automatically-assigned AOIs are not necessarily the optimal choice. As a consequence, an operator can change the assigned AOI, and avoid threat areas by changing a vehicle's goal or adding a waypoint to the path of the vehicle in order to go around the threat area.

Lastly, the control panel provides vehicle health information, as well as information on the vehicle's mission. The timeline displays the estimated time of arrival to waypoints and AOIs. Beneath the timeline is a mission progress bar that shows the amount of time remaining in the total simulation. Conceptually, the two interfaces depicted in Figures 3.3 and 3.5 are the same, however, the RESCHU interface is a much more realistic simulation environment than the one used in the previous experiment, and more directly tied to the DoD vision of one operator controlling multiple unmanned vehicles of various types.

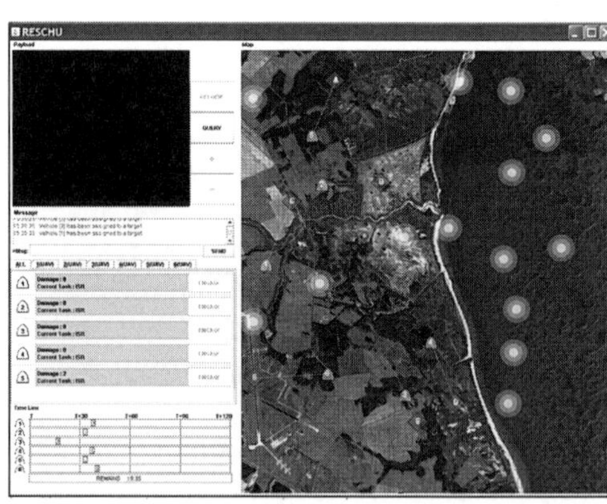

Figure 3.5 RESCHU interface

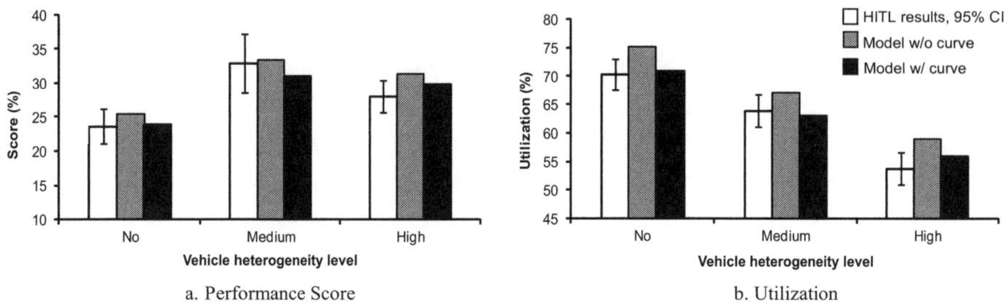

Figure 3.6 Human-in-the-loop (HITL) results versus model predictions per level of heterogeneity

Participants and Procedure

Given the relatively small numbers of participants in the previous experiment, we wanted to ensure that any significant trends resulting from this experiment had higher statistical power. As a result, seventy-four participants, six females and 68 males, between the ages of 18–50 completed the experiment. The participant who scored the highest received a $200 gift certificate. The experiment was conducted online with an interactive tutorial and an open-ended practice session. The website was password protected and participation was via invitation. All data were recorded to an online database. Demographic information was collected via a questionnaire presented before the tutorial. After participants felt comfortable, they could end the practice session and start the ten-minute experimental session.

Participants were instructed to maximize their score by 1) avoiding dynamic threat areas, 2) completing the maximum number of search tasks correctly, 3) minimizing vehicle travel times between AOIs, and 4) ensuring a vehicle was always assigned to an AOI whenever possible. They controlled five vehicles each, of mixed vehicle types, which will be detailed in the next section.

Experimental Design and Independent Variables

The experiment was a completely randomized design with vehicle team heterogeneity level as a between-subject condition: none (n = 26), medium (n = 25), and high (n = 23). The no heterogeneity condition composed of five MALE UAVs. The medium heterogeneous level had three MALE UAVs and two UUVs. Because the UUVs were slower than UAVs, they produced events less frequently. HALE UAVs visited targets that required identification, and once identified, either MALE UAVs or UUVs were assigned to the targets. Because the UUVs were slower than UAVs and the HALE UAVs did not have an associated visual task, the no heterogeneity condition composed of five MALE UAVs was the highest tempo scenario, (meaning events arrived more frequently) followed by the medium and then the high heterogeneity conditions. Such variations in team composition helped to determine how robust the model is, as the vehicles in the first experiment were a homogeneous set.

As in the previous experiment, the variables of interest for evaluating model predictions were score and operator utilization. The performance score was calculated as the proportion of the number of targets correctly identified normalized by the number of all possible targets that could have been identified. Operator utilization was calculated as in the previous experiment.

Results and Discussion

The observed average performance scores and utilizations for all three vehicle heterogeneity levels are presented in Figure 3.6, along with the predictions of the model with and without the workload-performance curve. The curve used to penalize operators by adding delay times for high and low utilizations was based on the distributions derived from the experimental data.

As seen in the previous experiment, the inclusion of the inverted U-shaped workload-performance curve allows the model's predictions to fall within the 95 percent confidence interval across all three different combinations of vehicle types. The DES model without the curve consistently over-predicted both performance scores and utilizations, and thus was significantly improved with the addition of the workload-performance model. These results, in combination with the previous experimental results, provide clear evidence that a parabolic workload-performance curve is critical in DES modeling efforts for human operator interaction with supervisory control systems.

Regarding the shape of the inverted U curve seen in the human-in-the-loop experiment, the distribution of the performance scores in 10 percent intervals (Figure 3.7) across all aggregated trials mirrors the hypothesized inverted-U relationship such that over- or under-utilization caused mission performance to degrade. Pairwise comparisons for Figure 3.7 revealed that 60–70 percent utilization corresponded to significantly higher scores than the majority of other utilization values ($p \leq .02$ across the various categories). In addition, 80–90 percent utilization resulted in lower performance scores than both 70–80 percent ($p = .03$) and 40–50 percent, ($p = .04$) utilization. Previous studies have also shown that when the operators work beyond 70 percent utilization, their performance degrades significantly (Schmidt 1978, Cummings and Guerlain 2007), so these results confirm these previous results that there can be a threshold for performance in terms of operator utilization.

Given that the performance score (the percentage of possible targets acquired) was context-specific, another dependent variable that is both more generalizable and more indicative of the possible negative impacts of performance problems is the delay time caused by operator inattention. This was discussed previously as delay times induced into the system by the loss of operator situation awareness, and these are the source of delay penalties described earlier.

Previous research has shown that these delays caused by operator inattention can account for the largest part of vehicle wait time, and significantly reduce the overall number of vehicles that a single operator could control (Cummings and Mitchell 2008). Since delay times caused by an operator who has lost situation awareness are a more objective measure of performance than a derived performance score, we examined how these wait times mapped to operator utilizations.

When these wait times caused by operator loss of situation awareness are mapped to the 10 percent utilization bins (Figure 3.8) (causing a U-shaped curve due to negative consequences of wait times as opposed to the inverted U-shaped positive outcome of performance score), the results differ somewhat from those in Figure 3.7. In the case of those operators responsible for the two UUVs, two MALE UAVs, and single HALE UAV (high heterogeneity), there is a clear skew to the higher utilization bins. This UV team was more cognitively complex to control than the no-homogeneity team, which as seen in Figure 3.8, produced a workload-performance curve more in keeping with the notional U-shaped function (inverted) in Figure 3.2. The medium level of heterogeneity operators experienced slight skews to the higher utilization bins, which suggests that for this experiment, as the complexity of the task increased, workload also increased.

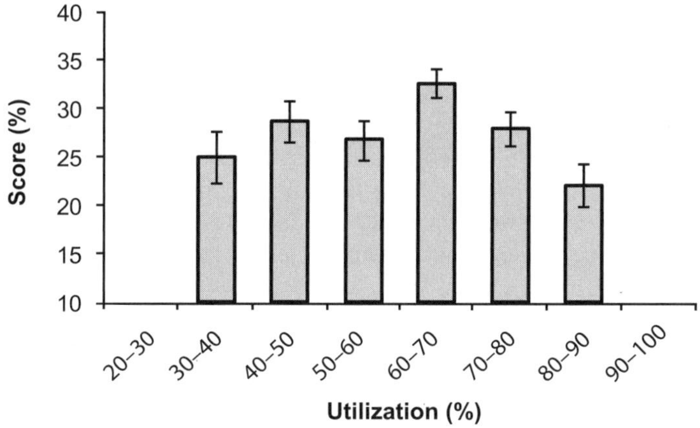

Figure 3.7 Experimental results for score versus utilization (with standard error bars)

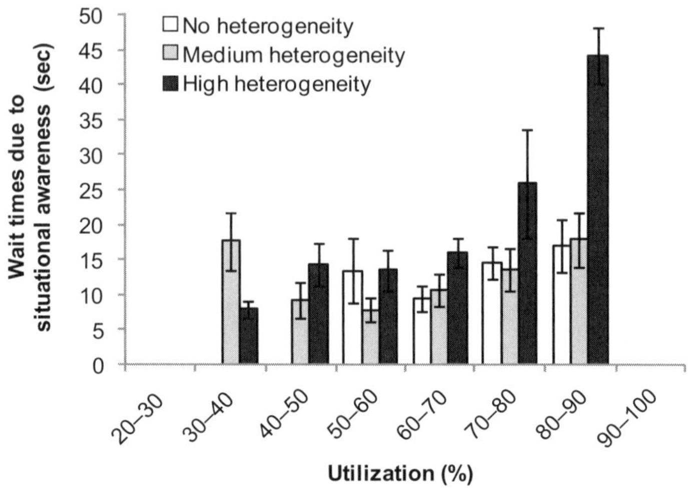

Figure 3.8 **The effects of increasing utilizations for induced delay times across the three levels of UT team heterogeneity**

Conclusions

With the voluminous increase in incoming information in network centric operations, identifying plateaus of optimal performance as a function of workload, as well as critical thresholds for degraded performance are key to designing systems that operators can effectively manage. Predicting regions of optimal performance or points of cognitive saturation is difficult for NCO systems because they are inherently task dependent, with potential varying levels of automation, operational tempos, training, experience, etc. Thus it is critical to develop simulation models for these futuristic systems that allow for exploration of these different design parameters without conducting extensive experimentation. However, these simulation models need accurate representations of human performance. In this paper, we have demonstrated that such a model is achievable through a discrete event simulation model, but that representing the negative impact of operator performance as a function of workload in a parabolic form is pivotal in improving this model's predictions.

While this work has generated the first known empirical evidence that some form of a parabolic workload-performance curve, suggested by the Yerkes-Dodson relationship, is both observable and when replicated in a quantitative model, adds significant value in human supervisory control discrete event simulation models of unmanned vehicles, this effort should still be considered preliminary. There are a number of related issues that require further investigation.

First, as suggested by Figure 3.8 and the results of the first experiment, true workload-performance curves are likely not parabolic in the symmetrical sense.

As seen in the second experiment, increasingly complex mixtures of vehicle teams caused different overall utilizations and correlated performances. Thus the nature of the task could dramatically vary an accurate workload-performance representation. However, in any modeling effort, it is not possible to always capture every variable influencing a system, thus if more general workload-performance curves could be found for task types, these relationships could be very valuable to human-system modeling efforts.

Another issue is the temporal and dynamic nature of utilizations. All utilizations reported here were post-hoc aggregate measures, which are not accurate for any instantaneous measure of workload. Thus more research is need to determine how utilizations can be measured in an online fashion, and moreover, what thresholds truly indicate poor performance—for example, does a person need to work at or above 70 percent utilization for some period of time before the negative effects are seen in human and/or system performance.

Another related area that requires further investigation is the rate of onset of high workload or utilization periods. The true measure of the impact of workload on performance may not be sustained utilization, but rather the onset rates of increasing utilizations. The other side of the curve represents cognitive underload, and the nature of sustained and variable rates of utilization changes also deserves further scrutiny.

Such investigations will be crucial in both aiding supervisory control modeling efforts, but they are also potentially valuable in the field of dynamic, adaptive automation design. If successful performance models of over or under cognitive load based on utilization can be developed, then more reliable forms of adaptive automation can be developed such that automation can intervene or assist human operators when a transition into a negative workload-performance region occurs.

Lastly, more fundamental work is needed to determine the nature of underlying physiologic or cognitive and psychological factors in contributing to performance decrements when operators move beyond some plateau. Previous research by Hancock and Warm (1989) suggests that operators seek a zone of maximal psychological capability that is constrained by their physiologic capability, and that once these zones are exceeded, performance degradation can occur. More work is needed to determine how psychological or physiologic processes are similar or different under high and low utilizations. Furthermore, future efforts are needed to determine if these zones can be generalized in such a way to include them in a quantitative model, in order to replicate observed human behavior so that the models can be used to predict both human and system performance.

References

Carbonell, J.R., Ward, J.L. and Senders, J.W. (1968) "A Queuing Model of Visual Sampling Experimental Validation", *IEEE Transactions on Man Machine Systems*, 9, 82–7.

Crandall, J.W. and Cummings, M.L. (2007a) "Identifying Predictive Metrics for Supervisory Control of Multiple Robots", *IEEE Transactions on Robotics—Special Issue on Human-Robot Interaction*, 23, 942–51.

Crandall, J.W. and Cummings, M.L. (2007b) "Attention Allocation Efficiency in Human-UV Teams", *AIAA Infotech@Aerospace Conference*, Rohnert Park, CA.

Crandall, J. and Cummings, M.L. (2007c) "Developing Performance Metrics for the Supervisory Control of Multiple Robots", *Human Robotics Interaction 2007* Conference, Washington DC.

Cummings, M.L., Bruni, S., Mercier, S. and Mitchell, P.J. (2007) "Automation Architecture for Single Operator, Multiple UAV Command and Control", *The International Command and Control Journal*, 1, 1–24.

Cummings, M.L. and Guerlain, S. (2007) "Developing Operator Capacity Estimates for Supervisory Control of Autonomous Vehicles", *Human Factors*, 49, 1–15.

Cummings, M.L. and Mitchell, P.J. (2008) "Predicting Controller Capacity in Supervisory Control of Multiple UAVs", *IEEE Transactions on Systems, Man, and Cybernetics—Part A: Systems and Humans*, 38, 451–60.

Department of Defense (2001) "Network Centric Warfare: Department of Defense Report to Congress", Office of the Secretary of Defense, Washington DC.

Department Of Defense (2007) "Unmanned systems roadmap", Office of the Secretary of Defense, Washington, DC, 2007–32.

Donmez, B., Nehme, C., Cummings, M.L., and de Jong, P. (in review) "Modeling Situational Awareness in a Multiple Unmanned Vehicle Supervisory Control Discrete Event Simulation", *IEEE Systems, Man, and Cybernetics—Part A: Systems and Humans*.

Hancock, P.A. and Ganey, H.C.N. (2003) "From the inverted-U to the extended-U: The Evolution of a Law of Psychology", *Human Performance in Extreme Environments*, 7, 5–14.

Hancock, P.A. and Warm, J.S. (1989) "A Dynamic Model of Stress and Sustained Attention", *Human Factors*, 31, 519–37.

Hebb, D.O. (1955) "Drives and the C.N.S. (Conceptual Nervous System)", *Psychological Review*, 62, 243–54.

Law, A.M. and Kelton, W.D. (2000) *Simulation Modeling and Analysis*, 3rd ed. (Boston: McGraw-Hill International Series).

Liu, Y. (1997) "Queuing Network Modeling of Human Performance of Concurrent Spatial and Verbal Tasks", *IEEE Transactions on Systems, Man, and Cybernetics*, 27, 195–207.

Liu, Y., Feyen, R. and Tshimhoni, O. (2006) "Queuing Network-model Human Processor (QN-MHP): a Computational Architecture for Multitask Performance in Human-Machine Systems", *ACM Transactions on Computer-Human Interaction*, 13, 37–70.

Nehme, C. Crandall, J.W. and Cummings, M.L. (2008) "Using Discrete Event Simulation to Model Situational Awareness of Unmanned-Vehicle Operators", *Modeling, Analysis and Simulation Center Capstone Conference*, Norfolk, VA.

Nehme, C.E. (2009) "Modeling Human Supervisory Control in Heterogeneous Unmanned Vehicle Systems", in *Aeronautics and Astronautics. vol. Doctor of Philosophy* (Cambridge, MA: Massachusetts Institute of Technology).

Pina, P.E., Cummings, M.L. Crandall, J.W. and Della Penna, M. (2008) "Identifying Generalizable Metric Classes to Evaluate Human-Robot Teams", *Metrics for Human-Robot Interaction Workshop* at the *3rd Annual Conference on Human-Robot Interaction*, Amsterdam, Netherlands.

Pinedo, M. (2002) *Scheduling: Theory, Algorithms, and Systems*, 2nd ed. (Englewood Cliffs, NJ: Prentice Hall).

Rouse, W.B. (1983) *Systems Engineering Models of Human-Machine Interaction*. (New York: North Holland).

Schmidt, D.K. (1978) "A Queuing Analysis of the Air Traffic Controller's Workload", *IEEE Transactions on Systems, Man, and Cybernetics*, 8, 492–8.

Senders, J.W. (1964) "The Human Operator as a Monitor and Controller of Multidegree of Freedom Systems", *IEEE Transactions on Human Factors in Electronics*, 5, 2–5.

Sheridan, T.B. (1969) "On How Often the Supervisor Should Sample", *IEEE Transactions on Systems, Science, and Cybernetics*, 6, 140–45.

Sheridan, T.B. (1992) *Telerobotics, Automation, and Human Supervisory Control*, (Cambridge, MA: The MIT Press).

Smith, P., McCoy, E. and Layton, C. (1997) "Brittleness in the Design of Cooperative Problem-Solving Systems: The Effects on User Performance", *IEEE Transactions on Systems, Man, and Cybernetics—Part A: Systems and Humans*, 27, 360–71.

Wickens, C.D. and Hollands, J.G. (2000) *Engineering Psychology and Human Performance*, 3rd ed. (Upper Saddle River, NJ: Prentice-Hall).

Wierwille, W.W. and Eggemeier, F.T. (1993) "Recommendations for Mental Workload Measurement in a Test and Evaluation Environment", *Human Factors*, 35, 263–81.

Yerkes R.M. and Dodson J.D. (1908) "The Relation of Strength of Stimulus to Rapidity of Habit-Formation", *Journal of Comparative Neurology and Psychology*, 18, 459–82.

Chapter 4
Systematic Measurements of Human Behavior in Naturalistic Settings

Matthew Rizzo

Joan Severson

Introduction

Accurate measurement of human activity is important to gauge the functional disability imposed by stress, fatigue, or injury and to monitor the outcome of interventions such as treatment or system redesign. Current tools such as observation in controlled laboratory and clinical settings or self-reported activity diaries and questionnaires may not reflect how humans act in real life, and may be inaccurate and misleading to supervisors and researchers. These inaccuracies may lead to improper actions, large sample sizes for research trials, and overall increased cost. Solutions to this problem can be derived through systematic behavioral and physiological measurements of humans while at work and play in naturalistic settings, taking advantage of great advances in sensor technology for simultaneously recording the movements of individuals and their internal states. This translational research motivates development of a system of "people tracker" tools for measuring human behavior and physiology in a wide range of geographical settings and disease states "in the wild." The multidisciplinary research needed to address these issues comprises scientists from the behavioral and social disciplines, with expertise in medicine, epidemiology, cognitive science, human factors engineering, computer science, and geography. This work benefits from the development of successful tracker prototypes and uses multiple integrated sensors and communication techniques to characterize subject movement and physiology, in synch with timestamps and specific geographic, environmental, and cultural settings. A basic implementation uses a wristwatch with GPS, accelerometers, and heart rate sensors linked with cell phone technology.

Background

People often act differently in controlled laboratory and clinical settings than they do in real life. Consequently, the goals, rewards, dangers, benefits, and time frames of sampled behavior can differ markedly between the clinic and "the wild."

A laboratory test may seem artificial or frustrating, and may not be taken seriously, resulting in a misleadingly poor performance. On the other hand, subjects may be on their best behavior and perform optimally when they know they are being graded in a laboratory or clinic, yet they may behave ineffectively in real life and fail to meet their apparent performance potentials at work, home, school, or in a host of instrumental activities of daily living. Solutions to these pitfalls in the study of brain—behavior relationships can be derived through rigorous observations of people at work and play in naturalistic settings, drawing from principles already being applied in studies of animal behavior and making use of great advances in sensor technology for simultaneously recording the movements of individuals, their surroundings, and their internal body and brain states.

In the absence of field observations, most of what we know about human behavior in the wild comes from human testimony (from structured and unstructured interviews and questionnaires) and epidemiology—a partial and sometimes inaccurate account of human behaviors that influence public health. Questionnaire tools may be painstakingly developed to assess all manner of behavioral issues and quality of life, and these generally consist of self-reports of subjects or reports by their family members, friends, supervisors, or caregivers. Incident reports of unsafe outcomes at work, in hospitals, or on roads completed by trained observers (for example, medical professionals, human factors experts, the police) are another source of similar information. However, these reports may be inaccurate because of poor observational skills or bias by untrained or trained observers. Subject reports may be affected by misperception, misunderstanding, deceit, and a variety of memory and cognitive impairments, including lack of self-awareness of acquired impairments caused by fatigue, drugs, aging, neurological or psychiatric disease, or systemic medical disorders. These information sources provide few data on human performance and physiology and often lack key details of what real people do in the real world.

In this vein, a study of cardiac rehabilitation measures after a myocardial infarction (for example, the amount a patient walked in six minutes) did not correlate well with patient activity on discharge (Jarvis and Janz 2005). Some of the individuals with the best scores in the hospital were among the least active at home and vice versa, raising concerns that patients with worse hearts were too active too soon, and that those with better hearts were returning to bad habits at home. Along similar lines, US soldiers addicted to heroin under stress in Vietnam often abstained without treatment back home, among family and work, and with the high cost of drugs.

Consider the challenge of assessing the real-world potential of individuals with decision-making impairments caused by brain damage, drugs, fatigue, and genetic or developmental disorders. Decision-making requires the evaluation of immediate and long-term consequences of planned actions, and it is often included with impulse control, insight, judgment, and planning under the rubric of executive functions (Benton 1991, Damasio 1996 and 1999, and Rolls 1999 and 2000). Impairments of these functions affect tactical and strategic decisions and actions in the real world.

Some of these impaired individuals have high IQs and perform remarkably well on cognitive tests, including specific tests of decision-making, yet fail repeatedly in real life. Some individuals who appear able to generate good plans may not enact these plans in the real world because they lack discipline, motivation, or social support, or choose alternative strategies with short-term benefits that are disadvantageous in the long term, as in the famous frontal lobe damage case of Phineas Gage and the well-described modern case of EVR (Damasio et al. 1994). Consequently, field observations using people tracking tools have the potential to provide unique information to fill in knowledge gaps on patient behavior, opening up completely new avenues and opportunities for research.

Much Human Behavior is Situational in Context

Behaviors depend critically on the environment in which they are observed. The idea that cognition must be studied in relation to real-world actions is embodied by the works of Piaget. His research on cognitive development in the infant and its dependence on exploring the environment anticipated the concept of situated-when-embodied cognition, as expressed by Clark (1997) in his book *Being There: Putting Brain, Body, and the World Together Again,* and by Hutchins (1995) in his book *Cognition in the Wild.*

Automated Behavior Monitors

Automated devices are widely used in laboratory neuroscience research for measuring general motor activity and tracking animals. People tracker tools build upon principles and methods from this research, typically involving automated recording of behaviorally relevant data from external devices or physiological sensors placed on or implanted within an animal. As technologies improve, applications to humans in real-world environments are expanding (Rizzo, Robinson, and Neale 2007). This research can draw from basic methods of ethology—the study of animal behavior in natural settings, pioneered by the Nobel Laureates Konrad Lorenz, Niko Tinbergen, and Karl von Frisch—that involve direct observation of behavior, including descriptive and quantitative methods for coding and recording behavioral events. Paper-and-pencil methods of the early ethologists have largely been replaced by the laptop computer and personal digital assistant (PDA)-based event-recording systems. It is now possible to make a continuous, real-time record of behavioral and physiological data, 24 hours a day, seven days a week, to discover the relationships between brain and behavior (Robinson et al. 1995, and Towell et al. 1987). The large volume of data forces researchers to adopt explicit strategies for sampling, summarizing, and simplifying the data to discover and extract useful information from it (for example, Anderson et al. 1998, and Waldrop 1990) as in the recent successful National Highway Traffic

Safety Administration (NHTSA)-sponsored study of driving performance, errors and crashes in 100 individuals, driving 100 total driver years in 100 instrumented vehicles (Neale et al. 2005). Moreover, such comprehensive behavioral data can now be accurately linked over extended periods to specific geographical locations, cultural environments, weather, and other exposures through links to a burgeoning set of GIS tools.

People Tracking

A people tracker system might be compared to a black box recorder in an aircraft or automobile, recording behavior sequences leading to an outcome or event of interest. This allows errors in a task or a pattern of unhealthy or risky behaviors to be measured, classified, and followed over time in wide-ranging interdisciplinary research studies of human behavior and health. The "people tracker" tools described here are designed to provide detailed information on the behavior, physiology, and pathophysiology of individuals in key, everyday situations, in settings, systems, and organizations where life goes on and things may go wrong. The research aims to measure more direct evidence of human behavior in the real world by combining sensors that capture human movement and physiological data in different, real-world locations in people who are seeing, feeling, attending, deciding, erring, and self-correcting during the activities of daily living.

Benefits of Collecting Data in Naturalistic Settings

It is becoming increasingly evident that data collection in a naturalistic setting is a unique source for obtaining critical human factors data relevant to behavior "in the wild." In pilot studies we have been addressing issues of device development, sensor choice, and placement, and needs for taxonomies for classifying devices and for classifying likely behavior from sensor outputs. To infer higher-level behaviors from sensor data in humans in naturalistic settings, it is also possible to apply ethological techniques that have been used to analyze behavior sequences in animals. These possibilities open large new areas of research to benefit human productivity, health, and organizational systems (Rizzo, Robinson, and Neale 2007). We also have extensive experience tracking human behavior over long distances using the instrumented vehicles ARGOS (Automobile for Research in Ergonomics and Neuroscience) and NIRVANA (Nissan Iowa Research Vehicle for Advanced Neuroergonomic Assessment).

We developed these mobile laboratories for measuring real-world behavior in drivers at risk for unsafe behavior owing to cognitive decline from neurological disorders, including traumatic brain injury, stroke, neurodegenerative disorders such as Alzheimer's and Parkinson's diseases, and advanced aging. We have extensive experience with processing of continuous signals from a large data stream

Figure 4.1 A subject in the instrumented vehicle, NIRVANA

(Uc et al. 2005, Anderson et al. 2007, and Dawson et al. 2009). Our experience with naturalistic data collection and analysis enabled us to develop a primary iteration of a people tracker. Our efforts are aimed at simple, reliable, and unobtrusive systems that are comfortable to wear and easy to use for researchers. Our initial development has resulted in a prototype system that has been used for beta testing and data analysis. The existing hardware and software system integrates wearable video, accelerometers, GPS, and heart-rate monitors. Data are synchronized and collected via wireless PDA for post-processed, web-based viewing and analysis of human behavior (Figure 4.2).

Specific aims are to (1) develop people-tracker software, networking, and hardware, (2) develop behavior analysis algorithms, (3) corroborate links between clinical data and people tracker data, (4) acquire data over an extended period, and (5) develop repositories of real-world, people tracker-derived behavioral data, accessible by investigators for a range of future hypothesis-driven research projects. Corroboration of the links between clinical measures and behavior (including extended recordings of, say, one month) can rely on pilot studies in individuals who are at particular risk for decreasing activity due to stress, fatigue and possible injury (for example, due to blast or chemical exposure in deployed soldiers). The tools developed for these purposes can be used to measure the behavior of people who are seeing, feeling, attending, deciding, erring, and self-correcting during activities of daily living over extended time frames. We expect that these tools will enable future multidisciplinary research to link human health and disease to real-world behavior, environmental exposures, and contexts in unprecedented ways, enabling the tracking of disease improvement or progression, and objectively assessing outcomes of interventions (drugs, devices, and rehabilitation) in military research and in NIH-sponsored clinical trials.

Software and Networking

A key component of this research is developing a software and network infrastructure to acquire, organize, store, and retrieve large volumes of data generated by people tracker tools. To effectively collect data in naturalistic settings, the network must be robust and discrete, allowing subjects to engage in their everyday activities without interference from the equipment while still providing complete data. The multidisciplinary nature of the research requires that the storage system allow multiple researchers with different data requirements easy and quick access to appropriate data. The overall system is based on an operational framework for people tracker data collection processing, integration, analysis, and reporting. Wearable physiologic sensors communicate wirelessly with, for example, a cell phone with GPS. Data adaptors accommodate to various interfaces to send information to a data repository, comprising raw and processed data from sensors. These data are further processed, allowing report generation and Web interface for user input, report, data query and analysis interfaces. A reasoning engine infers patterns of behavior and health and is attached to a geospatial database and logical database. This information can be linked to additional databases for weather, genetics, health records, demographics, epidemiology and others.

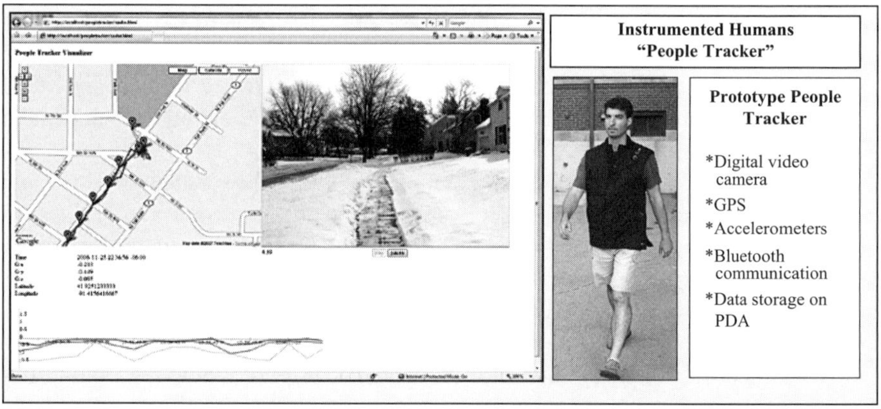

Figure 4.2 Prototype people tracker and data review tool

Right panel depicts a subject wearing the initial prototype tracker, which used a vest. Data are processed for analysis via Google Maps API, Macromedia Flash and JAVA. Left Panel: The graph shows raw data from tracker accelerometers in another subject wearing the tracker. The map locates where the subject was walking at specific times. Tracker video shows the snowy path ahead of the subject. No video data are collected with the basic extended use version of the tracker, which is embedded in a wristwatch.

Hardware Needs

The ability to track individuals in real world settings requires development of basic configurations and combinations of sensors, creating simple, rugged and unobtrusive sensor combinations that are comfortable to wear and easy to use. Guided by research needs, people tracker systems can be developed from a range of possible commercial off-the-shelf (COTS) sensor combinations. Briefly, these systems can be classified as inside-inside, inside-outside, and outside-inside (Mulder 1994). *Inside-inside systems* employ sensors and sources located on a person's body (for example, sensors to record heart rate). *Inside-outside systems* employ sensors on the body that detect external sources, such as accelerometers attached to the trunk or limbs of a person moving in the earth's gravitational field in order to track distance traveled and energy expended, as in pedometers. *Outside-inside systems* use external sensors that process various sources of information, such as a global positioning system (GPS) and Wi-Fi positioning systems (WPS) to process data from GPS and/or WPS sensors carried by or attached to a person. The basic people tracker aims to integrate data from all three perspectives.

Basic Tracker Configuration

The initial tracker builds on our existing prototype, employing an unobtrusive, wrist-mounted heart rate monitor to link physiology with real-world behavior, GPS placement of a subject at a specific location during events, and accelerometers for assessing/measuring/indexing body movement and energy expenditure. The basic tracker uses a waterproof wristwatch with integrated GPS, accelerometers, and heart rate sensors. The "watch" can be worn continuously for monitoring over prolonged periods. Data from accelerometers in the watch can function as an actigraphic index during sleep (Katayama 2001, and van Someren 1997). Data storage capacity currently allows at least 72 hours of continuous recordings from the sensors.

Data Downloading

A Bluetooth link to a cell phone, coupled to the watch, allows automatic downloading of stored data from the watch, transmission, and linkage to other databases, including geographic databases. Recorded data can be downloaded and transmitted daily (or with greater or lesser frequency depending on research needs). The cell phone linked with the watch has backup GPS and accelerometers for subject tracking. Both the watch and phone can be remotely tested by the investigators at any time to probe the integrity of the sensors and data collection. The cell phone is worn and removed as needed, does not need to be continuously near the watch, and can download data from the watch from up to 10 meters away.

GPS

Factors affecting the precision of coordinates relayed by GPS include receiver design, quality and shielding. Spatial resolution errors are generally 3 m or less outdoors, depending on proximity to physical features such as tall buildings or mountains, and 6 m or less indoors, depending on building materials and electronic interference (Nuckols et al. 2004). Resolution can be improved by augmenting GPS signals via nearby cell phone towers. Additional factors are accuracy of original base maps, including digitizing errors, time-lag in including roads and buildings in rapid growth areas, scale of experiments in relationship to available base maps (such as parcel data), and the error propagation arising from the need to re-project base maps into appropriate regional or national projections.

Tracker Quality Assurance and Modifications

Metrics for reliability and documentation of people tracker system failures, can use a root cause analysis process including failure mode and effects analysis, causal factor tree analysis, and fault tree analysis. Adjustments to hardware and software can be made as needed. A scalable software architecture can support system enhancements to address improvements in power supply for extended data collection capabilities, modular software adaptors to address evolving sensor API's (application programming interfaces), expanded use cases (for example, additional ways in which researchers will use people tracker systems) and database interfaces.

Future Designs

Development of subsequent people tracker system versions can be guided by functional requirements identified by focus group assessments and research priorities (for example, military and NIH). Technical considerations for future design include the lifecycle of available technologies and robustness of the hardware configurations. Depending on the research, these may require integrating body temperature, galvanic skin response (GSR), or other measures. Wi-Fi positioning systems can be added to generate signals that improve triangulation on subject position in enclosed spaces. They may also allow downloading of data directly form the sensors independently of the cell phone.

Behavior Analysis Algorithms

It is necessary to (a) determine how electronic outputs from the sensor array in subjects performing basic motor activities (such as walking, running, stair-climbing) compare to visual evidence of subject behavior obtained from miniature video cameras mounted on the subject and trained observers watching subjects in the field; and (b) develop and refine algorithms to process data from different versions of the people tracker and synch with other data sources (like a geographic information system—GIS) in order to track information such as distance traveled, places visited, most active periods, calories used, and heart beat range.

To accomplish (a) we compare the outputs of the prototype people tracker with corresponding visual evidence of subjects behaving in the field. Specifically, we compare observations from the accelerometers within the device against parallel sources of information—including synchronous miniature video cameras mounted to the subjects and trained on the environment, and observations by researchers trained in assessments of human movement—in order to determine the electronic "fingerprints" of key activity states (Ainsworth et al. 2000) such as walking, running, climbing stairs, traveling in a car, resting, and sleeping. The developed behavior analysis algorithm applications support recognition of performance "fingerprints." We are interested in validating signatures of key activities that may have prognostic value for health. This will allow us to classify likely behaviors from sensor output that can also serve as critical markers of success or decline in disease states. For example, it may turn out that climbing stairs is a key marker of successful overall mobility in the real world, for conditions such as knee arthritis or traumatic brain injury.

Moving from raw data to behavioral "fingerprints" requires that complex, time-varying data be reduced to identify the behavioral state of the person and parameters that quantify that state. For example, accelerometer data must be transformed into an estimate of whether the person is walking, sitting, or climbing stairs—and if the person is climbing stairs, how many flights and how quickly. Because no simple linear relationship links raw accelerometer data and behavioral state, a multi-stage data analysis approach is needed. The multi-stage data analysis begins by filtering and transforming the raw data. As an example, Fourier analysis provides a particularly useful transformation for characterizing behavior that involves periodic motion (McGehee et al. 2004). Walking is likely to produce a peak in the power spectrum at a frequency at approximately 1Hz that would not exist in the spectrum of a seated person. Substantial research has demonstrated the feasibility of this general approach, differentiating between walking and running with over 90 percent accuracy and even promising to differentiate people according to their particular walking style (Yam, Nixon, and Carter 2001).

Once the raw data are transformed into behaviorally relevant measures, the next stage identifies the behavioral state using Dynamic Bayesian Networks (DBNs). DBNs provide a promising framework to identify the person's state because they represent the conditional dependencies between variables and between variables over time. That is, the state of a node at time t can conditionally depend on the nodes in either t-1 or t time steps. The ability of DBNs to describe complex, non-linear, time-dependent relationships make them particularly applicable to human-behavior modeling; they have been used to detect affective state (Li and Ji 2005), fatigue (Ji, Zhu, and Lan 2004), lane change intent during driving (Kumagi and Akamatsu 2006) and driver cognitive distraction (Liang, Reyes, and Lee *in press*). For this application, the DBNs would take as input the transformed raw data and identify the behavioral fingerprint; for example, distinguishing between walking, running, and climbing stairs. The DBNs can be scaled to describe any number of distinct behaviors that might be of interest.

To accomplish (b) we focus on linking the output of the tracker with geographical information. The people tracker software analysis programs aim to track information such as distance traveled, places visited, most active periods, calories used, and heart beat range. These variables are recovered from integrated output from versions of the basic tracker and GIS databases. For example, by combining information from GPS on the tracker and regional GIS data, we can expand the details available for the variable, places visited, which can include different categories including residential area, commercial area, roadway, and even the specific address, depending on the needs of research. These measures can be used to characterize and analyze multiple dimensions of subject activity in cross-sectional assessments and longitudinal assessments of different cohorts (healthy and impaired). These details can be compared with subject reports obtained from standard tools such as the Life Space Questionnaire (Stalvey et al. 1999) to address higher-level issues such as self-awareness (or metacognition) and the accuracy of self-reporting research methods.

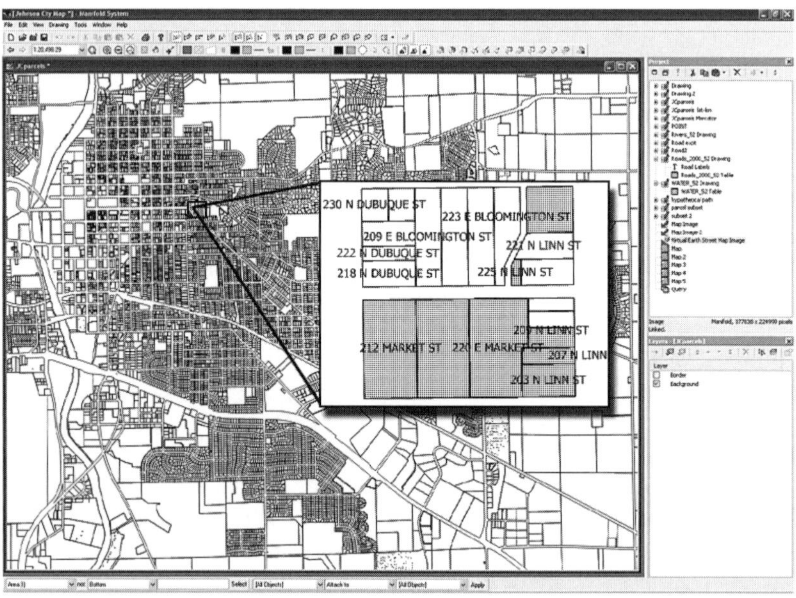

Figure 4.3 Sample of freely available Iowa City/Johnson County (GIS) parcel data (used in preliminary studies)

Geographic information system (GIS) data can be linked to a subject's time stamped biometric data, GPS coordinates, demographic and health records, and to geo-specific environmental data sources. In this inset commercial parcels are shaded and private parcels are not shaded.

Corroborating Links Between Clinical Data and People Tracker Data Collected in Different Cohorts

People tracker tools developed above can be used for locating and tracking human activity in a variety of cohorts. In medical research for example (Table 4.1), these tools can provide real-world determinations of disease magnitude, improvement or progression. The data could be used to test the efficacy of interventions (drugs, rehabilitation) in clinical trials that are relevant to the interests of several NIH institutes and the military.

Table 4.1 People tracker tasks

Examples of cohorts and conditions for study using people tracking tools	
1	Advanced age
2	Depression and other major psychiatric disorders
3	Stroke
4	Attention-deficit/hyperactivity disorder
5	Developmental Disorders such as, autism and Asperger's syndrome, and chronic childhood disease
6	Multiple sclerosis with remission, relapse, or progression
7	Head Injury
8	Alzheimer's disease and response to cholinesterase inhibitors and other medications
9	Parkinson's disease and response to dopaminergic agents or deep brain stimulation
10	Arthritis or fractures affecting the leg, hip, or spine, pre- and post-therapy
11	Alcohol or illicit drug use
12	Vestibular disease and other balance disorders
13	Pain syndromes (for example, back pain, migraine, fibromyalgia)
14	Chronic medical disorders (for example, cardiac, renal, hepatic, infectious)
15	Prescription drug use, for example, in cardiac disease, pulmonary disease, allergies, insomnia, and hypertension, and Schedule II and Schedule III drugs
16	Various genetic and epidemiologic cohorts

The utility of the people tracker tools can be demonstrated in tasks under differing levels of experimental control and environmental context, moving from the laboratory to the field. For instance, as a subject can perform a task over a period of about an hour in a task performed in a real world setting, or monitored for extended, continuous observations in the field over the course of a month outside of the view of the investigators. The former tasks can include a Baseline Mobility Assessment in which a subject is instructed to walk for six minutes around an uncluttered area of, say, a shopping mall (a) at comfortable rate, and (b) as briskly as possible without feeling uncomfortable or unstable. Subjects also (c) climb a set of stairs. Subjects may also perform a Navigation/Route Following task in which they must follow a route from a starting point to a target, designated by verbal instruction (resembling a set of verbal directions to a tourist), from memory and using a map. These different tasks depend on auditory comprehension, working memory, imagery, spatial abilities, planning and sequencing, and self-correction of errors to accomplish the task efficiently and avoid getting lost. These can be, (a) from verbal instruction to criterion: The subject is read a series of instructions and must to recite the directions twice correctly before setting out; (b) from being shown the route before: The research assistant takes the subject from a start point through a pre-scripted route to a target. The subject is taken back to the starting point and asked to repeat the trip; or (c) from a map of the environment: The subject holds a paper map and must use it to find a destination in the hospital. Dependent measures for these tasks are shown in Table 4.2.

Table 4.2 Tracking over an extended period

Dependent Measures in People Tracker Tasks
Total time to finish task
Distance traveled
Number of steps
Total time spent stopped
Total time spent moving
Total number of stops
Number of times revisiting or crossing the same place.
Total energy expenditure (based on distance traveled, weight, and BMI)
"Holding on" behaviors (hands on walls, railings, other people)
Use of cane or assistive device
Getting lost

These recordings demonstrate how people tracker tools provide unique objective evidence of real-world behavior in key study cohorts, including longitudinal assessments of individuals lasting up to one month. A *basic tracker for extended use* consists of a waterproof wristwatch with integrated GPS, accelerometers, and heart rate sensors and data storage capacity for at least 72 hours of continuous recordings. The watch is worn continuously for the entire month of this experiment and is safe to use while showering or bathing, but may be removed at any time if there is discomfort. The Bluetooth link to a cell phone and an accompanying the watch allows automatic downloading of data from the watch and transmission and linkage to the people tracker research databases (see later). These data are downloaded and transmitted daily. The cell phone linked with the watch is carried throughout the day by the subject and has backup GPS and accelerometers for subject tracking. Divergence of data from the cell phone and watch can flag the data stream for further inspection. Both the watch and phone can be called remotely as needed by the investigators to probe the integrity of the sensors and data collection. The cell phone will be worn on a location on the body core (for example, clipped to a belt), can be removed as needed, and does not need to be continuously near the watch. Data from accelerometers in the watch provide an actigraphic index (Katayama 2001, and van Someren 1997) during sleep. The extended tracker tool does not use video.

Geographic Information Systems (GIS)

Data from the extended tracker tool are subsequently linked to GIS data, as with our successful prototype people tracker. GIS are a burgeoning set of software tools for acquiring, storing, retrieving, analyzing, and displaying spatial data that greatly increase the use of data from the people tracker by placing these data in geographical and cultural context. GIS tools are now enabling researchers in epidemiology, medicine, medical geography, and public health to conduct unprecedented studies of disease exposure, agent source and distribution, disease clusters, health services access, health outcomes, and other key public health and health policy issues (Cromley 2003, McLafferty 2003, and Nuckols et al. 2004). In the past, GIS systems and analyses have not had access to continuous and objective data on the location and movement patterns of individuals (Nuckols et al. 2004); such information would provide novel evidence to link disease agents or exposures with health outcomes.

GIS-based analyses benefit greatly from mobile GPS to track individual humans (Nuckols et al. 2004, Rodriguez, Brown, and Troped 2005, and Shoval and Isaacson 2006). Much of the base spatial information (such as county parcel data and addressing) is freely available (see Figure 4.3). Cell phone GPS receiver and unobtrusive heart rate monitor data can be downloaded via Bluetooth or a physical connection to database software (either free software like PostgreSQL + PostGIS or commercial software, for example, Oracle Spatial). Each coordinate relayed to the database has a time stamp and a heart rate. For each coordinate, the GIS database tool attaches a location (address) that can be linked to a residence, business place, public place, or road (as in Figures 4.2 and 4.3). From the basic heart rate, time, and location data, people tracker algorithms will determine time spent in particular locations, mode of transportation (walk, drive), general activity level, and activity patterns. Thus GPS, heart rate data, and accelerometer data on the extended-use tracker are linked with GIS to recover places visited and when, social settings of those places, and types of transport (walk, run, car). Knowing how far and where people go and stop during a day and over weeks gets directly at the concept of life space, currently indexed by questionnaires or diaries.

Table 4.3 Measures for extended tracking

Dependent Measures
Time spent at home
Time sleeping
Time not moving, but awake
Time and distance traveling in a car
Number of episodes of climbing stairs
Time and distance walking
Time at work, in stores, etc.
Time exercising
Number of visits to a store, repeated trips in the same day
Total distance and radius of travel
Independent Measures
Time of day
Weather (precipitation, temperature, humidity, barometric pressure)
Season
Location: rural or urban; residential, commercial, or industrial; recreational, workplace, home

Privacy

No video is included in the extended use of the tracker. No video exists to link the person to the sensor outputs or derived behaviors. Any information linking the sensor data to a patient identification is encrypted and available only to the investigators. Sensor output is not linked to geographic information until post-processing. These data are also protected via encrypted links available only to the investigators. Subjects may drop out at any time and request that their data be deleted.

Experience Sampling Method (ESM)

Measurements of movement, physiology and behavior in naturalistic settings described above can be combined with the Experience Sampling Method (ESM) pioneered by Csikszentmihalyi and others. ESM tools attempt to validly describe variations in self-reports of mental processes and behavior in real world settings over time. ESM provides data on frequency, patterns and intensity of daily activity, social interaction, and movement; psychological (emotional, cognitive) and conative dimensions (mental processes or experiences geared toward action, such as impulse, desire, volition, and striving), and thoughts, including quality and intensity of thought disturbance (Csikszentmihalyi and Larson 1987). This approach can be applied to study a range of issues in normal and clinical populations, including in military settings.

Several studies provide evidence for validity by of ESM by showing correlations between ESM measures and physiological measures, baseline psychological tests and behavioral indices (Hekter, Schmidt, and Csikszentmihalyi 2007). In general, subjects respond to queries when alerted, up to several times per day, over varying periods of time (up to weeks) with the researcher not present. In these settings, tools are needed for alerting a participant, delivering queries and capturing participant responses. Triggered alerts may be random, timed, event based, or driven by user behavior, and delivered via cell phones, "personal digital assistants" (PDAs), pagers, watches or custom devices. Alerts may be as simple as an auditory beep or haptic buzz or include questions, delivered in audio or text form (live or scripted) via cell phones, PDAs, or paper, with responses in terms of yes/no, multiple choice, single or multi-answer, or freeform text or audio entry. Tools for capturing responses include cell phones, PDAs, audio recorders, and cameras (to depict a reported event or the context of a report).

ESM improves ecological validity by providing multiple assessments of behavior in context, in the real world, outside the laboratory. It reduces effects of intrusive observers, provides insights on behavior fluctuations over time (for example, Circadian, mood, alertness) and mitigates recall and memory biases by asking about current activity and mood. The behavioral data may reveal subconscious (covert) processes, provide a window to study activities and emotions often not accessible to observers, and over larger sample size and time interval. Individuals vary widely in their behavior over time. By allowing multiple observations within a person, ESM allows researchers to investigate variation over time and patterns of covariation between different variables.

Because ESM can investigate subjective appraisals and emotional reactions to stressful events in daily activities, it is relevant to civilian applications in healthcare and to military studies of acute stress disorder (ASD) and posttraumatic stress disorder (PTSD), which can negatively affect soldier readiness, deployability and safety. Altered stress sensitivity is independent of neurocognitive impairments (another risk factor for increased stress sensitivity). Exposures to severe stress (such as in the battlefield) may increase sensitivity to small stresses in daily life, giving rise to lasting behavioral lability (Myin-Germeys et al. 2009).

ESM challenges are that it may be time-consuming and demanding for participants, as well as more intrusive than one-time interviews and surveys. The participant must carry tools at all times and queries must not annoy subjects or influence experience, behavior and reports. Query frequency should fit temporal dynamics of processes of interest. Compliance is not guaranteed, as subjects are behaving in the field, without researcher guidance. Certain events may only show up with large samples sizes/durations. Power consumption, mentioned earlier, and service availability are also issues.

References

Ainsworth, B.E., Haskell, W.L., Whitt, M.C., Irwin, M.L., Swartz, A.M., Strath, S.J., O'Brien, W.L., Bassett, D.R., Jr., Schmitz, K.H., Emplaincourt, P.O., Jacobs, D.R., Jr. and Leon, A.S. (2000) "Compendium of Physical Activities: An Update of Activity Codes and Met Intensities", *Med Sci Sports Exerc*, 32, S498–504.

Anderson, C.M., Mandell, A.J., Selz, K.A., Terry, L.M., Wong, C.H., Robinson, S.R., Robertson, S.S. and Smotherman, W.P. (1998) "The Development of Nuchal Atonia Associated with Active (REM) Sleep in Fetal Sheep: Presence of Recurrent Fractal Organization", *Brain Res*, 787, 351–7.

Anderson, S.W., Rizzo, M., Skaar, N., Stierman, L., Cavaco, S., Damasio, H., and Dawson, J. (2007) "Amnesia and Driving", *Journal of Clinical and Experimental Neuropsychology*, 29, 1–12.

Benton, A. (1991) "The Prefrontal Region: Its Early History" in *Frontal Lobe Function and Dysfunction*, Levin, H., Eisenberg, H. and Benton, A. (eds), (New York: Oxford University Press), 3–12.

Clark, A. (1997) *Being There: Putting Brain, Body, and the World Together Again*, (Cambridge, MA: MIT Press).

Cromley, E.K. (2003) "GIS and Disease", *Annu Rev Public Health*, 24, 7–24.

Csikszentmihalyi, M. and Larson, R. (1987) "Validity and Reliability of the Experience-Sampling Method", *J Nerv Ment Dis*, 175, 526–36.

Damasio, A. (1999) *The Feeling of What Happens: Body and Emotion in the Making of Consciousness*, (New York, NY: Harcourt Brace & Co).

Damasio, A.R. (1996) "The Somatic Marker Hypothesis and the Possible Functions of the Prefrontal Cortex", *Philos Trans R Soc Lond B Biol Sci*, 351, 1413–20.

Damasio, H., Grabowski, T., Frank, R., Galaburda, A.M. and Damasio, A.R. (1994) "The Return of Phineas Gage: Clues About the Brain from the Skull of a Famous Patient", *Science*, 264, 1102–5.

Dawson, J.D., Anderson, S.W., Uc, E.Y., Dastrup, E. and Rizzo, M. (2009) "Predictors of Driving Safety in Early Alzheimer Disease", *Neurology*, 72, 521–7.

Hekter, J., Schmidt, J. and Csikszentmihalyi, M. (2007) *Experience Sampling Method: Measuring the Quality of Everyday Life*, (Thousand Oaks, CA: Sage Publications).

Hutchins, E. (1995) *Cognition in the Wild*, (ed), (Cambridge, MA: MIT Press).

Jarvis, R. and Janz, K. (2005) "An Assessment of Daily Physical Activity in Individuals with Chronic Heart Failure", *Medicine & Science in Sports & Exercise*, 37(5)S, S323–S324.

Ji, Q., Zhu, Z. and Lan, P. (2004) "Real-Time Nonintrusive Monitoring and Prediction of Driver Fatigue", *IEEE Transactions on Vehicle Technology*, 53, 1052–68.

Katayama, S. (2001) "Actigraph Analysis of Diurnal Motor Fluctuations During Dopamine Agonist Therapy", *Eur Neurol*, 46 Suppl 1, 11–17.

Kumagi, T. and Akamatsu, M. (2006) "Prediction of Human Driving Behavior Using Dynamic Bayesian Networks", *IEICE Transactions on Information and Systems*, E89–D, 857–860.

Li, X. and Ji, Q. (2005) "Active Affective State Detection and User Assistance with Dynamic Bayesian Networks", *IEEE Transactions on Systems, Man, and Cybernetic—Part A: Systems and Humans*, 35, 93–105.

Liang, Y., Lee, J. and Reyes, M. (2007) "Nonintrusive Detection of Driver Cognitive Distraction in Real Time Using Bayesian Networks", *Transportation Research Record*, 2018, 1–8.

McGehee, D., Lee, J., Rizzo, M., Dawson, J. and Bateman, K. (2004) "Quantitative Analysis of Steering Adaptation on a High Performance Driving Simulator", *Transportation Research Part F: Traffic Psychology and Behavior*, 7, 181–96.

McLafferty, S.L. (2003) "Gis and Health Care", *Annu Rev Public Health*, 24, 25–42.

Mulder, S. (1994) "Human Movement Tracking Technology." Hand Centered Studies of Human Movement Project, (Simon Fraser University) 94–101.

Myin-Germeys, I., Oorschot, M., Collip, D., Lataster, J., Delespaul, P. and van Os, J. (2009) "Experience Sampling Research in Psychopathology: Opening the Black Box of Daily Life", *Psychol Med*, 1–15.

Neale, V., Dingus, T., Klauer, S., Sudweeks, J. and Goodman, M. (2005) "An Overview of the 100-Car Naturalistic Study and Findings", International Technical Conference on the Enhanced Safety of Vehicles, Proceedings (CD-ROM), Washington, DC, 2005: National Highway Traffic Safety Administration.

Nuckols, J.R., Ward, M.H. and Jarup, L. (2004) "Using Geographic Information Systems for Exposure Assessment in Environmental Epidemiology Studies", *Environ Health Perspect*, 112, 1007–15.

Rizzo, M., Robinson, S.R. and Neale, V. (2007) "The Brain in the Wild" in *Neuroergonomics: Brain and Behavior at Work*, Parasuraman, R. and Rizzo, M. (eds), (Oxford, England: Oxford University Press),

Robinson, S.R., Wong, C.H., Robertson, S.S., Nathanielsz, P.W. and Smotherman, W.P. (1995) "Behavioral Responses of the Chronically Instrumented Sheep Fetus to Chemosensory Stimuli Presented *in Utero*", *Behav Neurosci*, 109, 551–62.

Rodriguez, D.A., Brown, A.L. and Troped, P.J. (2005) "Portable Global Positioning Units to Complement Accelerometry-Based Physical Activity Monitors", *Med Sci Sports Exerc*, 37, S572–581.

Rolls, E. (1999) *The Brain and Emotion*, (Oxford, England: Oxford University Press).

Rolls, E.T. (2000) "The Orbitofrontal Cortex and Reward", *Cereb Cortex*, 10, 284–94.

Shoval, N. and Isaacson, M. (2006) "Application of Tracking Technologies to the Study of Pedestrian Spatial Behavior", *Professional Geographer*, 58, 172–83.

Stalvey, B., Owsley, C., Sloane, M. and Ball, K. (1999) "The Life Space Questionnaire: A Measure of the Extent of Mobility of Older Adults", *Journal of Applied Gerontology*, 18, 479–98.

Towell, M.E., Figueroa, J., Markowitz, S., Elias, B. and Nathanielsz, P. (1987) "The Effect of Mild Hypoxemia Maintained for Twenty-Four Hours on Maternal and Fetal Glucose, Lactate, Cortisol, and Arginine Vasopressin in Pregnant Sheep at 122 to 139 Days' Gestation", *Am J Obstet Gynecol*, 157, 1550–57.

Uc, E.Y., Rizzo, M., Anderson, S.W., Shi, Q. and Dawson, J.D. (2005) "Driver Landmark and Traffic Sign Identification in Early Alzheimer's Disease", *J Neurol Neurosurg Psychiatry*, 76, 764–8.

Van Someren, E.J. (1997) "Actigraphic Monitoring of Movement and Rest-Activity Rhythms in Aging, Alzheimer's Disease, and Parkinson's Disease", *IEEE Trans Rehabil Eng*, 5, 394–8.

Waldrop, M.M. (1990) "Learning to Drink from a Fire Hose", *Science*, 248, 674–5.

Yam, C., Nixon, M. and Carter, J. (2001) "Extended Model-Based Automatic Gait Recognition of Walking and Running", Proceedings of the Third International Conference on Audio- and Video-Based Biometric Person Authentication, 278–83.

Chapter 5
Noninvasive Monitoring of Brain Function with Near Infrared Light

Andrew K. Dunn

Introduction

Noninvasive imaging with near infrared (NIR) light holds great promise for imaging in the brain in both animals and humans. NIR light has the ability to penetrate through several centimeters of tissue due to the relatively small absorption of the primary chromophores and low scattering in tissue in the wavelength range of 650–950 nm (Figure 5.1). Useful information can be extracted from NIR measurements due to the unique spectral absorption characteristics of biological chromophores as well as to exogenous contrast agents. The basis for most noninvasive optical measurements of brain function is the wavelength dependent absorption spectrum of the oxygenated and deoxygenated forms of hemoglobin. By measuring the intensity of NIR light that has penetrated the skull, interacted with cortical tissue, and scattered back through the skull, useful information about cortical activity can be quantified. In order to separate changes in oxy-hemoglobin and deoxy-hemoglobin, measurements must be performed at more than one wavelength in a manner that is exactly analogous to pulse oximetry.

Optical tomography and NIR spectroscopy (NIRS) have been used in a wide variety of *in vivo* noninvasive brain applications, due to their ability to monitor oxy- and deoxy-hemoglobin concentrations in applications such as functional activation (Villringer and Chance 1997, Chance et al. 1998, Franceschini et al. 2000, Gratton et al. 2000, and Toronov et al. 2000), and brain pathologies such as stroke and trauma (Robertson, Gopinath, and Chance 1995, and Hintz et al. 1999). In addition, both functional activation studies (Hock et al. 1996) and spectroscopic measurements (Hanlon et al. 1999) have been conducted in Alzheimer's patients. Another area where NIR measurements have proven to be particularly effective is in monitoring of cerebral hemodynamics in infants and children under both normal and pathological conditions (Franceschini et al. 2007, and Baird et al. 2002). Studies of brain function in children using traditional neuroimaging methods such as fMRI are often prohibitively difficult due to safety concerns. Since NIR methods employ low optical powers, they are considered to be extremely safe and as a result, have become widely used in experimental studies of normal brain development in children. Furthermore, infants often have the distinct advantage that their skulls are thinner than adults which makes light penetration more effective and leads to improved signal to noise ratios. In addition, many infants have little or no hair, which also improves coupling of light into the tissue.

NIR Measurement Techniques

NIR measurement configurations can be classified into three categories: time-domain, frequency-domain and continuous wave. In continuous wave (CW) measurements, the steady state intensity of NIR light that has been scattered throughout the tissue is measured, and these measurements are typically performed at two to four different wavelengths in the NIR region. CW measurements are the simplest and most common type of measurement since the instrumentation is straightforward and relatively inexpensive. Light sources are traditionally based on laser diodes which are readily available at wavelengths between 650 and 950 nm with sufficient powers to provide enough light to overcome the scattering losses in tissue. More recently, high power light emitting diodes (LEDs) have become available and these LEDs have the potential to reduce the cost of NIRS instruments and allow high-density arrays to be constructed (Zeff et al. 2007). Detectors typically consist of avalanche photodiodes (APD) which are sensitive in the NIR region and have high bandwidth. In most CW instruments, multiple light sources at two to four different wavelengths are each modulated at slightly different frequencies, and all light sources are turned on simultaneously. Each detector collects signals from multiple light sources simultaneously and the detector signals

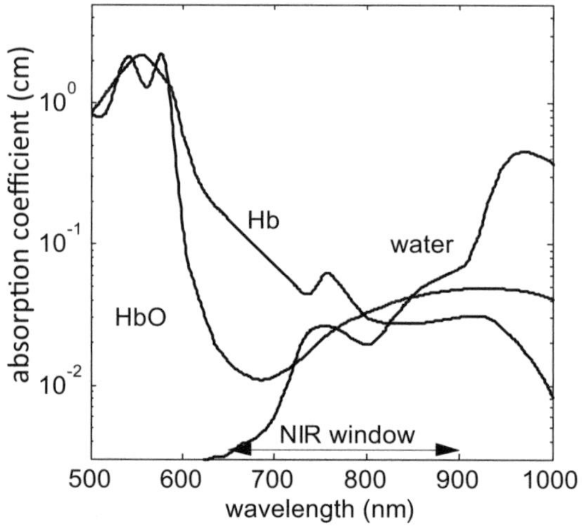

Figure 5.1 Absorption spectrum of biological chromophores; oxyhemoglobin (HbO), deoxyhemoglobin (Hb) and water illustrating the near infrared window in tissue between 650 and 900 nm

are demodulated to extract the contributions of each source (Siegel, Marota, and Boas 1999). This scheme also has the distinct advantage that ambient light, such as indoor lighting, does not interfere with the measurements so that measurements can be performed with the lights on.

In time-domain measurements, a short pulse (10^{-11}–10^{-14} seconds) of light is launched into the tissue, and as the photons undergo multiple scattering, the laser pulse is broadened. The temporally broadened pulse is then detected with a fast detection system. Almost all time domain systems use expensive pulsed lasers as sources. These lasers can be mode-locked Ti:sapphire lasers which are very expensive and operate at a single wavelength with pulse widths of 100–200 fs at a repetition rate of 70–80 MHz. Pulsed diode lasers are also used and these have pulse widths of tens of picoseconds at repetition rates up to 80 MHz. Detectors for time domain measurements are also expensive and typically rely on time-correlated single photon counting (TCSPC) hardware since the temporal resolution must be sub-nanosecond. Although they are considerably more expensive and complex than CW instruments, time domain instruments have the advantage of enabling quantitative measurement of baseline optical properties and hemodynamics whereas CW instruments are limited to measurements of hemodynamic changes from some baseline state.

Like time domain instruments, frequency domain instruments are complex and expensive. In frequency-domain measurements, the intensity of the illumination source is modulated at megahertz frequencies and both the amplitude and phase shift of the reflected, transmitted or fluorescent light are detected. Due to the different paths that photons travel through the tissue, the light reaching the detector is modulated at the same frequency but has a phase shift and a different modulation depth than the incident light. Therefore, by measuring the phase shift and/or the modulation depth at multiple modulation frequencies, information that is equivalent to time domain measurements is obtained. In particular, absolute baseline optical properties can be determined (Tromberg et al. 1997, Fishkin et al. 1997, and Franceschini et al. 1997), which enables comparison of baseline cerebral hemodynamics across subjects. Frequency domain instrumentation is slightly less complex than time domain instrumentation since ordinary laser diodes can be used, but compared with CW instrumentation, frequency domain instruments are considerably more complex. Frequency domain instruments also share many common features with frequency domain fluorescence lifetime instruments.

Probes for NIR measurement of cerebral hemodynamics are an important component of all NIR instruments. In general the probes consist of an array of optical fibers that are positioned on the head. Some of the fibers are connected to laser diodes and deliver light to the head, while others collect the scattered light and direct the light to detectors. The geometry of the probe is critical in determining the sensitivity of the system to a particular area of tissue. One of the most important details of a probe is the separation distance of the source fibers and detector fibers. Very small separations lead to sampling of superficial tissues while larger separations (2–4 cm) lead to deeper sampling. Larger source detector separation distances also lead to lower signal levels due to scattering of light and this ultimately places an upper limit on the source-detector separation, and therefore, penetration depth.

The arrangement of fibers in a probe depends on the target tissue and application. Figure 5.2 illustrates two common configurations for noninvasive measurements of brain activity in humans. The picture on the left shows a typical configuration for whole head measurement of somatosensory function. The custom made cap contains 32 sources and 32 detectors that cover both the left and right somatosensory cortex with source detector separations of 2–4 cm. Such a configuration enables bilateral measurement of motor and sensory function when the subjects perform tasks such as finger tapping (Franceschini et al. 2003). The second picture illustrates a frequency domain NIRS probe that was used for measurements of baseline cerebral hemodynamics in normal infants (Franceschini et al. 2007). This probe consists of a linear arrangement of fibers containing 32 sources and four detectors. The source detector separations varied from 1 to 2.5 cm in this probe.

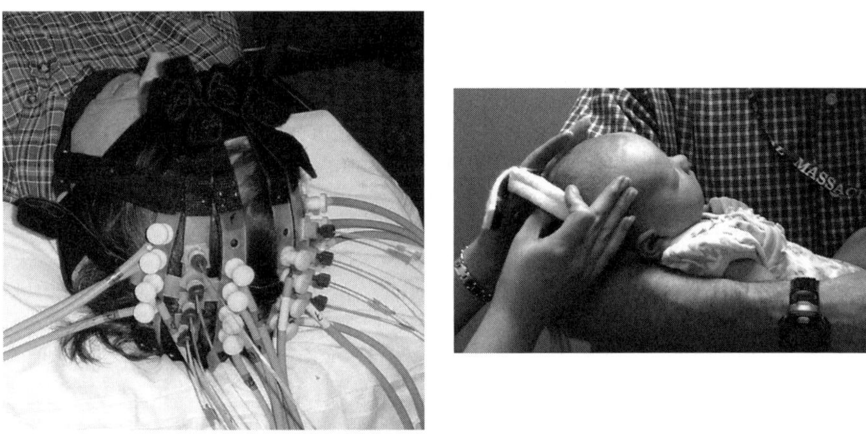

Figure 5.2 Examples of optical probes used in noninvasive NIR measurements in humans

The image on the left illustrates an array of optical fibers that cover both hemispheres while the image on the right illustrates a probe with a linear arrangement of fibers used for monitoring cerebral hemodynamics in infants.

Extraction of Physiological Information from NIR Measurements

All NIR measurements involve conversion of scattered light intensities to physiological parameters such as hemoglobin oxygenation or volume. Unlike x-ray CT where x-ray photons travel in straight lines, NIR photons travel highly tortuous paths due to light scattering. Therefore, conversion of NIR measurements to functional parameters is a complex process that requires solution of an ill-posed inverse problem (Arridge 1999). The primary distinction between NIRS and DOT lies in the manner in which the measurements are processed. NIRS measurements typically do not involve full image reconstruction and therefore, consist primarily of spatially integrated time courses of hemoglobin oxygenation and volume changes. The time courses are usually derived from nearest neighbor information. In other words, only adjacent source and detector measurement pairs are used in the conversion process. Although still complicated, NIRS inverse problems are considerably less complex than full image reconstruction with DOT.

DOT image reconstruction is done in many different ways using many different algorithms. The basic approach in an iterative image reconstruction is to assume a particular baseline hemodynamic state of the tissue and to use an appropriate model to predict a set of DOT measurements for each source-detector pair given the baseline hemodynamics and tissue geometry. The predicted values are then compared with the actual measurements and, based on the difference, a new set of spatially varying optical properties is generated. This iterative process continues until the difference between the measured and predicted values is minimized. The details of the various implementations of the iterative fitting are the subject of a large set of literature (see, for example, Arridge 1999).

One of the crucial aspects of this inverse problem in both NIRS and DOT measurements is developing accurate models of the forward problem. Since light propagation in tissues is complex due to scattering and absorption, most inverse problems have relied on simplifying assumptions for tissue geometry and model of light propagation. Typically the diffusion approximation to the radiative transport equation is used to predict light distribution in tissues since it yields analytical expressions that can be quickly evaluated. Monte Carlo methods on the other hand are not subject to these geometrical simplifications and can accurately model spatially varying absorption and scattering. This improved accuracy comes, however, at the cost of computational complexity. Most Monte Carlo simulations require from minutes to several hours of computation times, which has made them very difficult to use in inverse problems. Recently however, perturbation Monte Carlo methods have been developed which now permit the use of Monte Carlo models in iterative inverse solutions (Hayakawa et al. 2001).

The Monte Carlo method works by tracing individual photon paths through a 3-D tissue. Random numbers are used to sample probability distributions specifying the distance between scattering locations as well as the angular deviation of the photon at each scattering event. After tracing a sufficiently large number of photons (typically 10^6–10^8), accurate statistics about signal levels and light distributions within tissues can be developed. Figure 5.3 illustrates the distribution of light within a tissue for a typical source-detector pair in a NIRS or DOT measurement (Boas et al. 2002). Photons were launched into the tissue through a source fiber indicated by the down arrow. The Monte Carlo model computed the photons arriving at the detector as a function of time. Each image depicts the spatial distribution of photons arriving at the detector within a 50 picosecond (ps)-wide temporal window centered on a given time point (that is, 200±25 ps, 600±25 ps, and 1200±25 ps). This figure also illustrates the temporal dependence of the sampling depth on the detected intensity. The CW equivalent sampling distribution is also illustrated for comparison. These results illustrate how Monte Carlo simulations can be used to assess the spatial distribution of detected light.

The spatial resolution of NIR images is typically significantly lower than that of MRI, but NIR imaging provides unique physiological information because of its sensitivity to oxy- and deoxyhemoglobin as well as exogenous fluorophores. A multi-modality approach employing NIR and MR imaging can capitalize on the high spatial resolution of MRI and high specificity of optical tomography. Previous research has included simultaneous NIRS and fMRI of the hemodynamic response to functional activation in humans (Strangman et al. 2002), optical imaging and

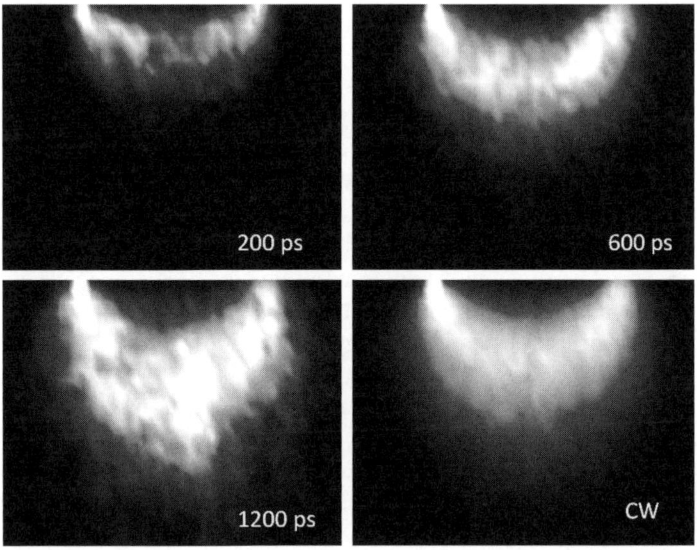

Figure 5.3 Illustration of the sampling functions for a typical source detector fiber pair in time resolved NIR imaging

The profiles were computed with Monte Carlo simulations and show the distribution of detected photons within a time window centered at 200, 600 or 1200 ps. The bottom right image shows the distribution for CW measurements.

MRI in rats (Pogue and Paulsen 1998), detection of breast tumors (Chang et al. 1997), and contrast enhanced MRI and optical tomography in human breast (Ntziachristos et al. 2000). Whereas many optical imaging reconstructions assume simplified geometries such as a planar, homogeneous semi-infinite medium, the anatomical structural information provided by the MRI can be used to guide the optical image reconstruction algorithm in a much more realistic manner (Pogue and Paulsen 1998, Ntziachristos et al. 2002b, and Schweiger and Arridge 1999). While simultaneous MRI and optical tomography is becoming more common, there is still a need for continued basic development of instrumentation and algorithms, particularly in applications involving small animals.

Figure 5.4 illustrates how anatomical MR scans can be incorporated into Monte Carlo simulations. In this example, an MR scan was segmented into four different tissue types (Boas et al. 2002). The Monte Carlo simulation then modeled the launching of photons at one point on the scalp, and quantified the spatial distribution of light within the head of those photons that would be collected by a detector fiber that is located three centimeters from the source fiber. The four different images illustrate the sensitivity profiles for CW, time domain and frequency domain measurements. In particular, the greater sampling depth and increased sensitivity to deeper structures of time domain measurements is evident when comparing the profiles in Figures 5.4c and 5.4d.

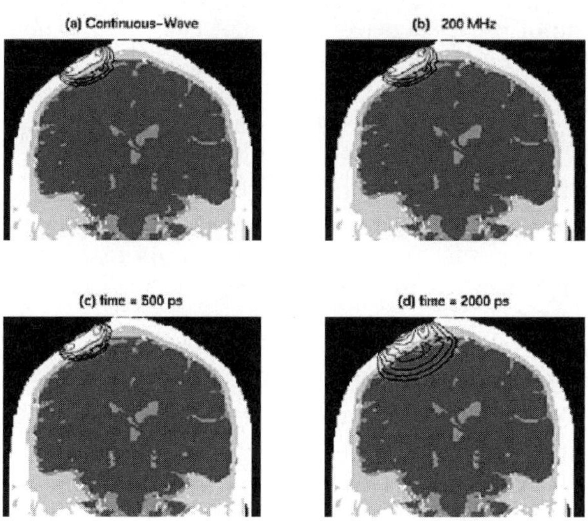

Figure 5.4 Examples of how Monte Carlo models can be used to predict the sensitivity profiles in the head by incorporating anatomical MRI information into the model (adapted from Boas et al. 2002)

Molecular Imaging

Much of the optical tomography to date has relied on differences in the intrinsic absorption and scattering properties between tissues as the primary source of contrast. However, a great deal of interest has been expressed recently in using fluorescent contrast agents with excitation and emission wavelengths in the NIR region that provide very high contrast between the tissue of interest and the background tissue (Klose and Hielscher 2003, Sevick-Muraca, Houston, and Gurfinkel 2002, Ramanujam et al. 1994, Godavarty et al. 2003, Jiang et al. 1998, and Gannot et al. 2003). Fluorescence based optical tomography provides spatial maps of both fluorophore concentration and lifetime, provided time-domain or frequency domain measurements are used. Many efforts have concentrated on using indocyanine green (ICG), a common NIR fluorophore, as a contrast mechanism, due to its fluorescence efficiency and the fact that it is FDA approved for use in humans. For example, Ntziachristos et al (2000) used ICG as a contrast agent in the breast during concurrent MRI and optical imaging, while Reynolds et al (1999) used ICG to image mammary tumors in an animal model, and Godavarty et al. (2003) used a frequency-domain fluorescence instrument to image a large tissue-simulating phantom. Others have imaged fluorescence lifetimes (O'Leary et al. 1996, and Sevick-Muraca et al. 1998) in both animal models and tissue simulating phantoms.

Some of the most exciting applications in fluorescence imaging have been in the development of probes targeted at the molecular level. Traditional imaging techniques such as x-ray CT and MRI rely primarily on intrinsic sources of contrast, which can limit their sensitivity. Until recently, detection and characterization of disease were based primarily on anatomic changes that were assumed to reflect the physiologic and molecular changes that actually underlie the particular pathology.

In vivo molecular imaging has attracted a great deal of attention in recent years because it relies on specific molecules as the source of image contrast, which can provide much higher specificity than traditional imaging techniques. The success of molecular imaging depends on several key elements: (a) the development of imaging probes able to overcome biological delivery barriers, (b) the use of amplification strategies, and (c) the availability of fast, sensitive, high-resolution imaging techniques (Weissleder and Ntziachristos 2003).

Examples of molecular imaging have included the use of bioluminescent indicators (Contag et al. 1998), NIR activatable probes, and NIR targeted contrast agents for tumor imaging (Achilefu et al. 2000). More recently, NIR fluorescent probes have been combined with tomographic imaging approaches to provide 3-D images in vivo (Ntziachristos et al. 2002a). The combination of targeted probes and fluorescence tomography will permit noninvasive, quantitative imaging studies to be undertaken, such as detection of Alzheimer's in both animal and human studies. Coupled with the advances in molecular imaging probes, there is a need for continued development of advanced instrumentation to capitalize on the potential for in vivo noninvasive detection of these probes. In particular, the ability to accurately quantify both fluorescence concentration and lifetimes requires a time-resolved system, or a frequency domain instrument, neither of which are commonly available.

As in NIRS and DOT measurements of intrinsic hemodynamic signals, noninvasive NIR imaging of fluorescent markers requires the solution of an inverse problem. Many approaches to fluorescence image reconstructions have been taken and each approach typically involves the solution of a forward problem at both the excitation and emission wavelengths. Figure 5.5 demonstrates the improvement spatial localization of time-domain image reconstructions that can be achieved with a priori knowledge of the fluorescence lifetimes (Kumar et al. 2005). In this example experimental measurements were taken in a tissue phantom containing two fluorophores as depicted in Figure 5.5a. When a traditional image reconstruction was performed (Figure 5.5b) the two fluorophores cannot be separated. However, when image reconstructions are performed of the components at specific fluorescence lifetimes, the two fluorophores can be separated (Kumar et al. 2005). This example is one of many in the rapidly emerging field of fluorescence molecular imaging.

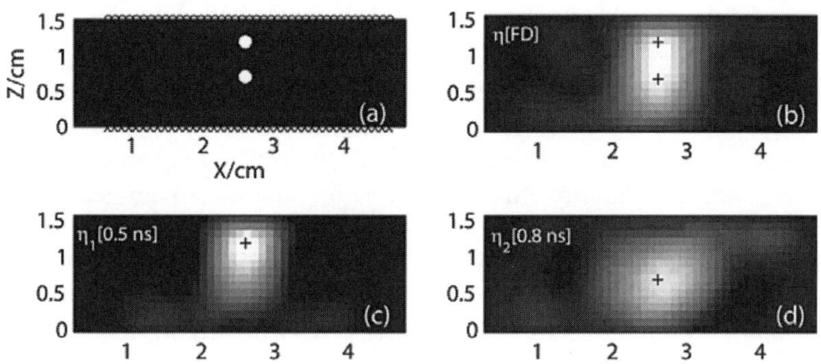

Figure 5.5 Demonstration of experimental fluorescence lifetime based tomography

(a) Measurement geometry with two fluorescent tubes located at depths of 1.2 and 0.7 cm. (b) Reconstructed fluorescence yield using traditional diffusion model. (c) and (d) Reconstructions of lifetime specific components enable improved spatial locatization of fluorophores (adapted from Kumar et al. 2005).

Conclusion

Noninvasive measurements of brain function using NIR light hold tremendous potential for many applications. NIR techniques can be used in situations where traditional neuroimaging methods are not feasible such as developmental studies in infants and children and continuous bedside monitoring in hospitals. Due to the compact instrumentation of CW NIRS instruments, they have the potential to be developed into compact portable devices that can be used to monitor cerebral function in dynamic, complex environments. Although more research must be performed in order to assess the feasibility of such studies, the large numbers of NIRS studies that have been performed have established the utility of NIRS as a reliable and robust method for monitoring cerebral hemodynamics.

References

Achilefu, S., Dorshow, R.B., Bugaj, J.E. and Rajagopalan, R. (2000) "Novel Receptor-Targeted Fluorescent Contrast Agents for in Vivo Tumor Imaging", *Invest Radiol*, 35, 479–85.

Arridge, S.R. (1999) "Optical Tomography in Medical Imaging", *Inverse Problems*, 15, R41–93.

Baird, A.A., Kagan, J., Gaudette, T., Walz, K.A., Hershlag, N. and Boas, D.A. (2002) "Frontal Lobe Activation During Object Permanence: Data from near-Infrared Spectroscopy", *Neuroimage*, 16, 1120–25.

Boas, D., Culver, J., Stott, J. and Dunn, A. (2002) "Three Dimensional Monte Carlo Code for Photon Migration through Complex Heterogeneous Media Including the Adult Human Head", *Opt Express*, 10, 159–70.

Chance, B., Anday, E., Nioka, S., Zhou, S., Hong, L., Worden, K., Li, C., Murray, T., Ovetsky, Y., Pidikiti, D. and Thomas, R. (1998) "A Novel Method for Fast Imaging of Brain Function, Non-Invasively, with Light", *Opt Express*, 2, 411–23.

Chang, J., Graber, H.L., Koo, P.C., Aronson, R., Barbour, S.L. and Barbour, R.L. (1997) "Optical Imaging of Anatomical Maps Derived from Magnetic Resonance Images Using Time-Independent Optical Sources", *IEEE Trans Med Imaging*, 16, 68–77.

Contag, P.R., Olomu, I.N., Stevenson, D.K. and Contag, C.H. (1998) "Bioluminescent Indicators in Living Mammals", *Nat Med*, 4, 245–7.

Fishkin, J.B., Coquoz, O., Anderson, E.R., Brenner, M. and Tromberg, B.J. (1997) "Frequency-Domain Photon Migration Measurements of Normal and Malignant Tissue Optical Properties in a Human Subject", *Appl Opt*, 36, 10–20.

Franceschini, M.A., Fantini, S., Thompson, J.H., Culver, J.P. and Boas, D.A. (2003) "Hemodynamic Evoked Response of the Sensorimotor Cortex Measured Noninvasively with near-Infrared Optical Imaging", *Psychophysiology*, 40, 548–60.

Franceschini, M.A., Moesta, K.T., Fantini, S., Gaida, G., Gratton, E., Jess, H., Mantulin, W.W., Seeber, M., Schlag, P.M. and Kaschke, M. (1997) "Frequency-Domain Techniques Enhance Optical Mammography: Initial Clinical Results", *Proc Natl Acad Sci USA*, 94, 6468–73.

Franceschini, M.A., Toronov, V., Filiaci, M., Gratton, E. and Fantini, S. (2000) "On-Line Optical Imaging of the Human Brain with 160-Ms Temporal Resolution", *Opt Express*, 6, 49–57.

Franceschini, M.A., Thaker, S., Themelis, G., Krishnamoorthy, K.K., Bortfeld, H., Diamond, S.G., Boas, D.A., Arvin, K. and Grant, P.E. (2007) "Assessment of Infant Brain Development with Frequency-Domain near-Infrared Spectroscopy", *Pediatr Res*, 61, 546–51.

Gannot, I., Garashi, A., Gannot, G., Chernomordik, V. and Gandjbakhche, A. (2003) "In Vivo Quantitative Three-Dimensional Localization of Tumor Labeled with Exogenous Specific Fluorescence Markers", *Appl Opt*, 42, 3073–80.

Godavarty, A., Eppstein, M.J., Zhang, C., Theru, S., Thompson, A.B., Gurfinkel, M. and Sevick-Muraca, E.M. (2003) "Fluorescence-Enhanced Optical Imaging in Large Tissue Volumes Using a Gain-Modulated Iccd Camera", *Phys Med Biol*, 48, 1701–20.

Gratton, G., Sarno, A., Maclin, E., Corballis, P.M. and Fabiani, M. (2000) "Toward Noninvasive 3-D Imaging of the Time Course of Cortical Activity: Investigation of the Depth of the Event-Related Optical Signal", *Neuroimage*, 11, 491–504.

Hanlon, E.B., Itzkan, I., Dasari, R.R., Feld, M.S., Ferrante, R.J., McKee, A.C., Lathi, D. and Kowall, N.W. (1999) "Near-Infrared Fluorescence Spectroscopy Detects Alzheimer's Disease in Vitro", *Photochem Photobiol*, 70, 236–42.

Hayakawa, C.K., Spanier, J., Bevilacqua, F., Dunn, A.K., You, J.S., Tromberg, B.J. and Venugopalan, V. (2001) "Perturbation Monte Carlo Methods to Solve Inverse Photon Migration Problems in Heterogeneous Tissues", *Opt Lett*, 26, 1335–7.

Hintz, S.R., Cheong, W.F., van Houten, J.P., Stevenson, D.K. and Benaron, D.A. (1999) "Bedside Imaging of Intracranial Hemorrhage in the Neonate Using Light: Comparison with Ultrasound, Computed Tomography, and Magnetic Resonance Imaging", *Pediatr Res*, 45, 54–9.

Hock, C., Villringer, K., Muller-Spahn, F., Hofmann, M., Schuh-Hofer, S., Heekeren, H., Wenzel, R., Dirnagl, U. and Villringer, A. (1996) "Near Infrared Spectroscopy in the Diagnosis of Alzheimer's Disease", *Ann NY Acad Sci*, 777, 22–9.

Jiang, H., Paulsen, K.D., Osterberg, U.L. and Patterson, M.S. (1998) "Frequency-Domain near-Infrared Photo Diffusion Imaging: Initial Evaluation in Multitarget Tissue-like Phantoms", *Med Phys*, 25, 183–93.

Klose, A.D. and Hielscher, A.H. (2003) "Fluorescence Tomography with Simulated Data Based on the Equation of Radiative Transfer", *Opt Lett*, 28, 1019–21.

Kumar, A.T., Skoch, J., Bacskai, B.J., Boas, D.A. and Dunn, A.K. (2005) "Fluorescence-Lifetime-Based Tomography for Turbid Media", *Opt Lett*, 30, 3347–9.

Ntziachristos, V., Tung, C.H., Bremer, C. and Weissleder, R. (2002a) "Fluorescence Molecular Tomography Resolves Protease Activity in Vivo", *Nat Med*, 8, 757–60.

Ntziachristos, V., Yodh, A.G., Schnall, M. and Chance, B. (2000) "Concurrent Mri and Diffuse Optical Tomography of Breast after Indocyanine Green Enhancement", *Proc Natl Acad Sci USA*, 97, 2767–72.

Ntziachristos, V., Yodh, A.G., Schnall, M.D. and Chance, B. (2002b) "Mri-Guided Diffuse Optical Spectroscopy of Malignant and Benign Breast Lesions", *Neoplasia*, 4, 347–54.

O'Leary, M.A., Boas, D.A., Li, X.D., Chance, B. and Yodh, A.G. (1996) "Fluorescence Lifetime Imaging in Turbid Media", *Optics Letters*, 21, 158–60.

Pogue, B.W. and Paulsen, K.D. (1998) "High-Resolution near-Infrared Tomographic Imaging Simulations of the Rat Cranium by Use of a Priori Magnetic Resonance Imaging Structural Information", *Opt Lett*, 23, 1716–18.

Ramanujam, N., Mitchell, M.F., Mahadevan, A., Warren, S., Thomsen, S., Silva, E. and Richards-Kortum, R. (1994) "In Vivo Diagnosis of Cervical Intraepithelial Neoplasia Using 337-Nm-Excited Laser-Induced Fluorescence", *Proc Natl Acad Sci USA*, 91, 10193–7.

Reynolds, J.S., Troy, T.L., Mayer, R.H., Thompson, A.B., Waters, D.J., Cornell, K.K., Snyder, P.W. and Sevick-Muraca, E.M. (1999) "Imaging of Spontaneous Canine Mammary Tumors Using Fluorescent Contrast Agents", *Photochem Photobiol*, 70, 87–94.

Robertson, C.S., Gopinath, S.P. and Chance, B. (1995) "A New Application for near-Infrared Spectroscopy: Detection of Delayed Intracranial Hematomas after Head Injury", *J Neurotrauma*, 12, 591–600.

Schweiger, M. and Arridge, S.R. (1999) "Optical Tomographic Reconstruction in a Complex Head Model Using a Priori Region Boundary Information", *Phys Med Biol*, 44, 2703–21.

Sevick-Muraca, E.M., Reynolds, J.S., Troy, T.L., Lopez, G. and Paithankar, D.Y. (1998) "Fluorescence Lifetime Spectroscopic Imaging with Measurements of Photon Migration", Ann NY Acad Sci, 838, 46–57.

Sevick-Muraca, E.M., Houston, J.P. and Gurfinkel, M. (2002) "Fluorescence-Enhanced, near Infrared Diagnostic Imaging with Contrast Agents", *Curr Opin Chem Biol*, 6, 642–50.

Siegel, A., Marota, J.J. and Boas, D. (1999) "Design and Evaluation of a Continuous-Wave Diffuse Optical Tomography System", *Opt Express*, 4, 287–98.

Strangman, G., Culver, J.P., Thompson, J.H. and Boas, D.A. (2002) "A Quantitative Comparison of Simultaneous Bold Fmri and Nirs Recordings During Functional Brain Activation", *Neuroimage*, 17, 719–31.

Toronov, V., Franceschini, M.A., Filiaci, M., Fantini, S., Wolf, M., Michalos, A. and Gratton, E. (2000) "Near-Infrared Study of Fluctuations in Cerebral Hemodynamics During Rest and Motor Stimulation: Temporal Analysis and Spatial Mapping", *Med Phys*, 27, 801–15.

Tromberg, B.J., Coquoz, O., Fishkin, J.B., Pham, T., Anderson, E.R., Butler, J., Cahn, M., Gross, J.D., Venugopalan, V. and Pham, D. (1997) "Non-Invasive Measurements of Breast Tissue Optical Properties Using Frequency-Domain Photon Migration", *Philos Trans R Soc Lond B Biol Sci*, 352, 661–8.

Villringer, A. and Chance, B. (1997) "Non-Invasive Optical Spectroscopy and Imaging of Human Brain Function", *Trends Neurosci*, 20, 435–42.

Weissleder, R. and Ntziachristos, V. (2003) "Shedding Light onto Live Molecular Targets", *Nat Med*, 9, 123–8.

Zeff, B.W., White, B.R., Dehghani, H., Schlaggar, B.L. and Culver, J.P. (2007) "Retinotopic Mapping of Adult Human Visual Cortex with High-Density Diffuse Optical Tomography", *Proc Natl Acad Sci USA*, 104, 12169–74.

PART 2
Cognition During Sleep Deprivation

Chapter 6

Individual Differences to Sleep Deprivation Vulnerability and the Neural Connection with Task Strategy, Metacognition, Visual Spatial Attention, and White Matter Differences

Matthew Rocklage

W. Todd Maddox

Logan T. Trujillo

David M. Schnyer

Introduction

Sleep deprivation (SD) is known to have disastrous real-world consequences due to its influence on mental functioning. In addition to its general effect on functioning, SD often elicits large individual differences in how persons function; however, why some individuals function at a continuously high rate over a long period without sleep while others experience a sharp dissolution of functioning after losing just a few hours of sleep remains a mystery. The differences that may be present between individuals and those factors that contribute to SD vulnerability are not well known.

Four separate approaches to understanding individual differences that contribute to SD vulnerability are introduced in this article. While these are by no means comprehensive studies, they serve to introduce a broad spectrum of approaches with some tantalizing preliminary results that may help make sense of differential vulnerability to SD. Three of these domains examine individual differences in cognitive functioning in the realm of decision-making strategies, metacognitive abilities, and visual-spatial attention. An additional area examines whether individual differences in brain white matter development are related to SD vulnerability. Though seemingly different, this broad spectrum of approaches to understanding SD vulnerability may reveal critical cognitive and/or neural factors that reflect multiple aspects of an individual's response to SD.

The Population

All of the work reported here was part of a US Army-funded project performed across a two-year period with young participants from The UT Austin, West Point Military Academy, and the US Army military base at Fort Hood, Texas. All sleep-deprived participants were tested when rested on Day 1 and then again 24 hours later after being kept awake continuously with light mental and physical activities between sessions. Participants were monitored at all times to ensure they remained awake. No prior record of sleep history was available, but the night before each session, participants were peer monitored in order to ensure a complete night's sleep.

Decision-Making Strategies during Category Learning

Category learning is fundamental to the survival of all organisms. When we judge a group of approaching individuals as friends or enemies we are categorizing. Two category learning domains that are of particular interest are rule-based and information-integration categories (Ashby et al. 1998, Ashby and Maddox 2005, Maddox and Ashby 2004). Rule-based categories are those for which the optimal decision rule is verbalizable. For example, if all members of one category are large, and all members of the other category are small, then the optimal strategy would be to determine the size of the target stimulus and to apply the following verbal rule: "if the stimulus is small place it in category A; if the stimulus is large place it in category B." Optimal rule-based category learning is thought to be mediated by a hypothesis-testing system that reasons in an explicit fashion, and is dependent on conscious awareness (Ashby et al. 1998; Ashby and Maddox 2005, Maddox and Ashby 2004, Nomura et al. 2007, Poldrack and Foerde 2008, Seger 2008). The hypothesis-testing system relies on working memory and executive attentional processes (Waldron and Ashby 2001, Zeithamova and Maddox 2006) and appears to depend on a neural circuit involving dorsolateral prefrontal cortex, anterior cingulate, the head of the caudate nucleus, and medial temporal lobe structures (Filoteo et al. 2005, Love and Gureckis 2007, Nomura et al. 2007, Seger and Cincotta 2005 and 2006).

Unlike rule-based categories, the optimal decision rule for information-integration categories is not verbalizable, but instead requires a pre-decisional integration of information from two or more stimulus dimensions (usually expressed in different physical units). An example of information-integration categories composed of circular sine-wave gratings is shown in Figure 6.1. The optimal strategy (denoted by the solid diagonal line) is not verbalizable because it involves a linear integration of information from dimensions expressed in incommensurable units (for example, orientation and spatial frequency). Optimal information-integration category learning is thought to be mediated by a procedural system (Ashby et al. 1998; Ashby and Maddox 2005, Maddox and Ashby 2004, Nomura et al. 2007, Poldrack and Foerde 2008, Seger 2008). Unlike the hypothesis-testing system, the procedural learning system is not consciously penetrable and instead operates by associating regions of perceptual space with actions that lead to reward. The procedural system is implemented by a circuit involving inferotemporal cortex and the posterior caudate nucleus (Ashby et al. 1998, Ashby and Ennis 2006, Nomura et al. 2007, Seger and Cincotta 2005, Wilson 1995).

One of the most well known and highly studied rule-based category learning tasks is the Wisconsin Card Sort Task (WCST; Heaton 1980). SD has been shown to adversely affect performance in the WCST. Specifically, Herscovitch, Stuss and Broughton (1980) found that sleep-deprived individuals were impaired in the WCST with the ratio of perseverative errors within a category to total perseverative errors increasing after SD. However, to our knowledge no studies have examined the effects of SD on information-integration categorization (although studies have examined SD effects on procedural motor learning tasks (Walker and Stickgold 2004), and none have examined these effects in US Military Academy cadets.

Metacognitive Abilities

The capacity to accurately reflect on one's performance and to then compensate for any perceived deficits is an important human trait. This capability to reflect on one's self and one's abilities is referred to as "metacognition." Despite metacognition's supposed connection with the prefrontal cortex (PFC; Schnyer, Nicholls, and Verfaellie 2005), which has been shown to be particularly susceptible to SD (Chee and Choo 2004), past research has indicated that there may be little change in a person's ability to know how well they are performing on that task under conditions

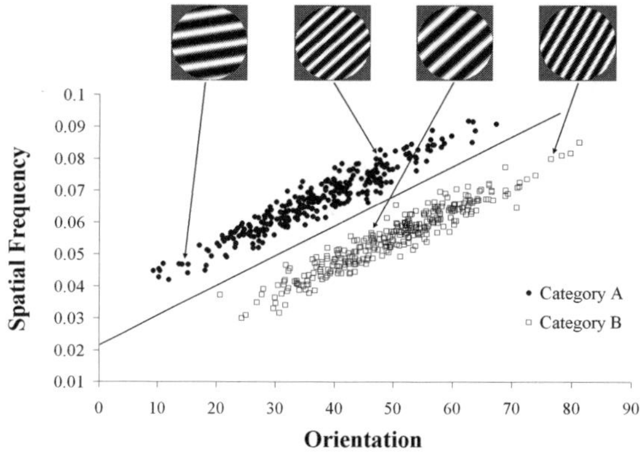

Figure 6.1 Scatterplot of the stimulus used in the experiment

The filled symbols denote stimuli from Category A and the open symbols denote stimuli from Category B. The diagonal line indicates the optimal decision bound. Four sample stimuli and their associated locations in the scatterplot are included.

of SD (Baranski 2007). A preliminary examination of this issue is presented in a second study where individual differences in metacognitive ability are examined. In this study, participants are asked to reflect on the difficulty of an integrated decision-making task on a trial-by-trial basis both when rested and again when sleep deprived. The accuracy of their judgments is assessed by comparing their perceived difficulty level with their actual performance.

ERP Examination of Visual-Spatial Attention

The ability to selectively attend to a single information source while excluding irrelevant items is crucial to many aspects of cognitive performance. Previous research has found SD to impair behavioral indices of selective visual-spatial attention (Gunter et al. 1987, Norton 1970, McCarthy and Waters 1997) and change the activity of brain structures meditating this important cognitive ability (Portas et al. 1998, Mander et al. 2008, Chee et al. 2008, Tomasi et al. 2009). In a recent study (Trujillo, Kornguth, and Schnyer, under review), examined the effects of SD on voluntary shifts of selective visual-spatial attention driven by factors endogenous to individuals versus attentional shifts driven primarily by exogenous factors. The study showed that as little as 24 hr of SD leads to a significant reduction in amplitude of the N1 event-related potential (ERP) response, a bioelectric measure of brain activity known to be highly sensitive to manipulations of attention (Luck, Woodman, and Vogel 2000) only when endogenous shifts of visual-spatial attention were required. In an additional analysis of that study presented here, how well SD-related reductions in the N1 ERP response predict decrements in an individual's behavioral indices of exogenous and endogenous selective attention was examined.

White Matter Differences

Individual differences in cognitive ability and their decline with aging or disease have been tied to brain structure across a multitude of domains (Assaf and Pasternak 2008, Jones et al. 1999, Medina et al. 2006). For example, using diffusion tensor magnetic resonance imaging (DTI) Madden et al. (2009) investigated the relationship between age-related declines in cognitive performance and fractional anisotropy (FA), a measure thought to reflect the number, myelination, and compactness of white matter (WM) tracts (Beaulieu, 2002). Specifically, lower FA values reflected a deficit in age-related decision-making ability seen in the older group. In SD research, failing to respond in a timely fashion to a stimulus due to "microsleeps" is called "lapsing" and is one such impairment that accompanies SD (Dinges et al. 1997). Here, in the final study, the findings from research were examined as to whether similar differences in the microstructure of WM tracts might correlate with cognitive functioning and impairment under conditions of SD across a group of healthy young individuals.

Methods and Results

Decision-Making Strategies in Category Learning.

Separate groups of US Military Academy cadets participated in both Control and Sleepless (sleep deprivation) conditions (n = 21 and 28, respectively). Participants completed five 100-trial blocks of category learning on Day 1. On each trial, participants were asked to categorize a single stimulus into one of two categories by pressing one of two keys on the keyboard and then received corrective feedback. Twenty-four hours after initial testing, participants in the Control and Sleepless test completed an additional three 100-trial blocks in the same category-learning task.

Performance was equivalent between the Sleepless and Control groups during Day 1 (see Figure 6.2, left hand graph). Performance during Day 2 was significantly worse for the Sleepless group relative to the Control group (see Figure 6.2, right hand graph). Thus, SD clearly led to a decline in information-integration performance. Finally, Day 2 performance was significantly better than Day 1 performance for the Control group suggesting that sleep consolidation effects do emerge in information-integration classification learning.

The accuracy-based analyses suggest that SD led to a performance decline during Day 2. However, it is well known that the use of information-integration versus rule-based strategies to solve information-integration categorization leads to large performance differences. Thus, it would be advantageous to determine whether the effects of SD differed as a function of the type of learning strategy that participants utilized on Day 1. To address this issue, we applied decision-bound

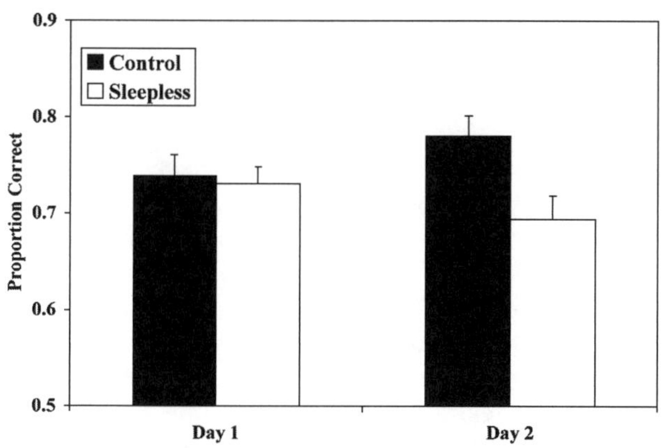

Figure 6.2 Proportion correct for the Control and Sleepless groups during Day 1 (final session) and Day 2 (overall)

Standard error bars are included.

models (Ashby and Gott 1988, and Maddox and Ashby 1993) to the final block of data from Day 1 that either assumed that the participant used an information-integration classification strategy or a rule-based classification strategy (see Maddox et al. *in press,* for details). Several comments are in order. First, Sleepless participants who used information-integration strategies performed better on Day 2 than those who used rule-based strategies. Second, Sleepless participants who used information-integration strategies held up better on Day 2 than Control participants who used rule-based strategies. Finally, Sleepless participants using information-integration strategies showed no significant performance drop on Day 2 relative to Day 1. The results of this study support the idea that individual differences in strategy implementation can determine one's ability to continue an optimal level of functioning under conditions of SD.

Metacognition

Twenty-six participants from the West Point US Military Academy (n = 5), Fort Hood Army installation (n = 16), and The University of Texas at Austin (n = 5) completed two sets of an integrative decision-making (ID) task, once before and once after 24 hours of total sleep deprivation (TSD).

During an fMRI experiment examining the effects of SD on neural functioning, participants engaged in a number of decision-making tasks. One required making perceptual matching decisions about black and white novel stimuli. Participants were asked to determine whether one and only one of two presented abstract stimuli matched a third "target" stimuli and respond "yes" if so and "no" to all other possibilities (no match and both match). In this way, participants were required to integrate their initial evaluation of each of the test stimuli with the final decision to respond "yes" or "no." This task has been referred to as an integrative decision-making task (ID). Interspersed with this and other decision-making tasks was a visual-motor control (VMC) task. The VMC task required participants to signify, by button press, on which side of the screen an "X" was presented. A percent change score was calculated for the VMC task ((Day 2 VMC accuracy—Day 1 VMC accuracy) / Day 1 VMC accuracy) to calculate participants' change in performance between days. Previous work has indicated that tasks like the VMC seem to be more sensitive in revealing the relationship between those more and less vulnerable to SD as they require only sustained attention to a simple task. Thus, cognitive vulnerability to SD may be best captured by a lower-level task that emphasizes primary attention (see Lim and Dinges 2008). Once finished with the decision-making task, participants engaged in a metacognitive judgment task where they rated the relative difficulty of 48 more ID trials. All participants performed both sets of tasks, the longer fMRI experiment, and the ID metacognitive judgment task when rested (Day 1) and after 24 hours of TSD (Day 2).

Groups were separated based on a median split of their percent change in performance on the VMC task. Results show that those more vulnerable to TSD (VUL group) and those less vulnerable to TSD (NON group) did not differ on how accurate they actually were on the ID task on both Day 1 and Day 2. By contrast, they did differ in the accuracy of judgments made about their own performance. Specifically, the VUL group had significantly greater insight into how well they were performing pre-TSD than the NON group. This difference disappears, however, on the Day 2 when the VUL group declines in its metacognitive ability while the NON group appears to show a numerical increase in the accuracy of these judgments (see Figure 6.3). Both this decline and increase in metacognitive ability between days, however, are non-significant on post-hoc testing.

Visual Spatial Attention

Fourteen West Point cadets (eight females and six males) and 12 Fort Hood soldiers (two females and ten males) performed modified Attentional Network Tasks (ANTs; Fan et al., 2007) that used exogenously and endogenously cued letter target stimuli to index brain networks underlying selective attention functions. A schematic of the basic ANT protocol is shown in Figure 6.4.

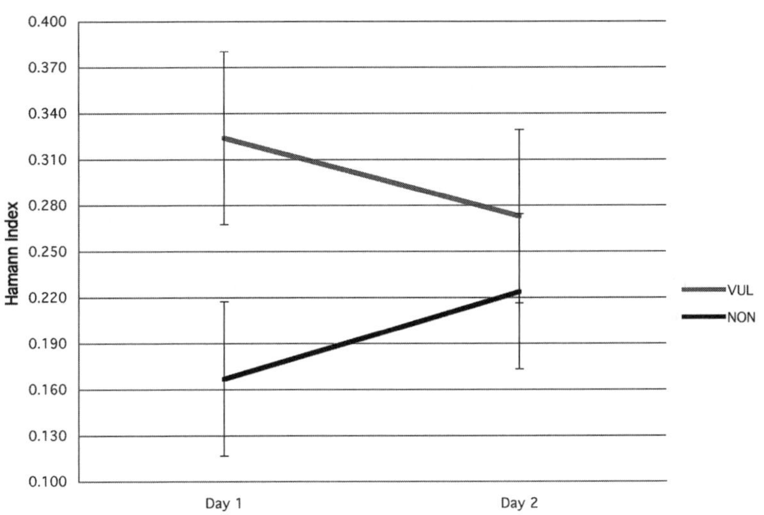

Figure 6.3 Hamann index for VUL and NON groups showing correlation between task accuracy and judgment accuracy on Day 1 and Day 2

Standard error bars are included. Hamann index ranges from -1 (complete disagreement between task accuracy and judgment) to 1 (complete agreement) and zero represents chance accuracy.

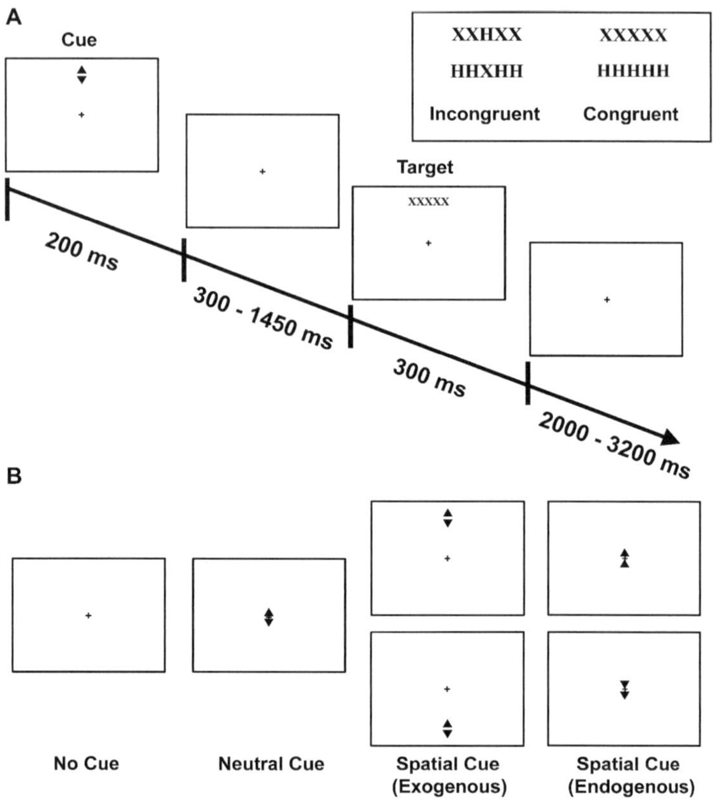

Figure 6.4 Basic Attention Network Test (ANT) protocol

Subjects categorized the center letter of a horizontal string of five letters as being the same as ("congruent") or different than ("incongruent") the flanking letters (Figure 6.4A, inset). Target trials were further divided according to whether the targets were preceded by a cueing stimulus (double triangle) that was predictive of the spatial location of the subsequently presented target. For no cue trials (Figure 6.4B, left column), no cueing stimulus preceded the targets. For neutral cue trials (Figure 6.4B, middle column), targets were preceded by a double triangle symbol that was non-predictive of the target's spatial location. For spatial cue trials (Figure 6.4B, right columns), targets were preceded by a double triangle symbol that predicted the subsequent spatial location of the targets with 100 percent accuracy. All cues were presented for 200 ms, while targets were on for 300 ms; response time limit was 2000 ms. Cue/target interstimulus intervals ranged from 300 to 1450 ms (M = 842 ms), while inter-trial intervals ranged from 2000 to 3200 (M = 2562 ms).

Subjects were administered two versions of this task that assessed either exogenous or endogenous allocation of selective attention. The exogenous and endogenous versions of the ANT were exactly the same except for the spatial cue condition. In the exogenous ANT (Figure 6.4B, third column), the spatial cueing triangle pointed in opposite directions, but appeared in the spatial locations of the subsequently presented targets. Thus, the exogenous ANT attentional shifts were driven primarily by a factor external to the subjects in that the onset of the cueing stimuli captured attention towards the location of the subsequently presented target. In the endogenous ANT (Figure 6.4B, fourth column), the spatial cueing triangles appeared at central fixation with triangles pointing in the direction of the spatial locations of the subsequently presented targets. Here, attentional shifts were driven primarily by a factor internal to the subjects in that they voluntarily shifted their attention after interpreting the triangle directions.

Participants performed the ANTs on two days separated by 24 hours of total sleeplessness. We previously reported (Trujillo, Kornguth, and Schnyer, under review) that overall response times (RTs) (collapsed across Spatial, Neutral, and No Cue trials conditions) significantly increased and that overall accuracy rates significantly decreased from Day 1 to Day 2 for both the Exogenous and Endogenous ANT (see Table 6.1).

Table 6.1 Mean accuracy rates (%) and RTs (ms) collapsed across cue conditions per task and per day. Standard error values in parentheses.

	Accuracy		RT	
Endogenous	95 (1)	92 (1)	718 (21)	795 (27)
Exogenous	96 (1)	91 (1)	735 (21)	798 (24)

ERPs were recorded while subjects performed the two ANTs. We found (Trujillo, Kornguth, and Schnyer, under review) overall N1 (defined as a nadir in the EEG trace recorded at 130–200 ms) ERP responses (again collapsed across all cue conditions) to be significantly reduced from Day 1 to Day 2 over dorsal parietal scalp locations for the Endogenous ANT (Figure 6.5, left panel), but not for the Exogenous ANT (Figure 6.5, right panel). We did not observe between-group differences for the behavioral or N1 measures.

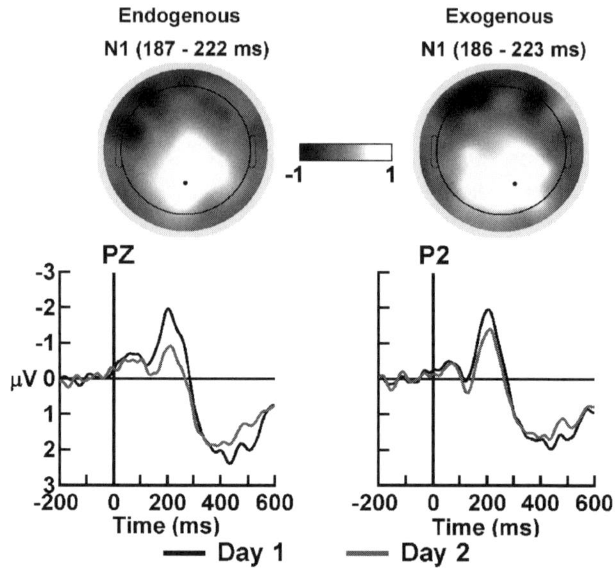

Figure 6.5 Representative grand-average target-locked N1 ERP responses collapsed across cue conditions for Day 1 (black line) and Day 2 (grey line)

Negative polarity is oriented upwards. Waveform scalp locations are indicated by black dots on topographic maps (top row) displaying mean Day 2—Day 1 differences over the stated intervals. Light colors indicate positive values, dark colors indicate negative values. Scalp map differences are presented in normalized dimensionless units.

Here, we report analyses of the relationship between SD-related reductions in the N1 ERP response of individuals and the behavioral differences observed during the Exogenous and Endogenous ANTs. We investigated this relationship by computing correlations between the Day 2—Day 1 differences in overall N1 ERP amplitude and the same differences in accuracy rates and RTs. As correlation analyses are severely affected by the presence of outliers, we removed subjects with N1, accuracy, or RT differences exceeding ±3 standard deviations of the group mean. This procedure resulted in the elimination of one subject from the Fort Hood group, bringing the total number of subjects for this analysis to 25.

Figure 6.6 plots individual participant Day 2—Day 1 N1 vs RT differences for the Endogenous (Figure 6, top panel) and Exogenous (Figure 6, bottom panel) ANTs. A significant correlation ($r = 0.52$, $p < 0.01$) was found for the Endogenous ANT. Since the N1 ERP component is negative in polarity, positive Day 2—Day 1 differences indicate smaller N1 responses on Day 2 vs Day 1. Thus, this positive correlation between N1 and RT across-day differences indicates that those participants with large N1 reductions from Day 1 to Day 2 exhibited larger RT increases from Day 1 to Day 2 than participants with small N1 reductions across the two days. In contrast, no significant relationship was observed between N1 and RT across-day changes for the Exogenous ANT (r = –0.08, $p > 0.72$). Furthermore, we found no significant correlations between N1 amplitude and accuracy rate changes across day for either task (Endogenous: r = –0.27, $p > 0.19$; Exogenous: r = –0.08, $p > 0.71$).

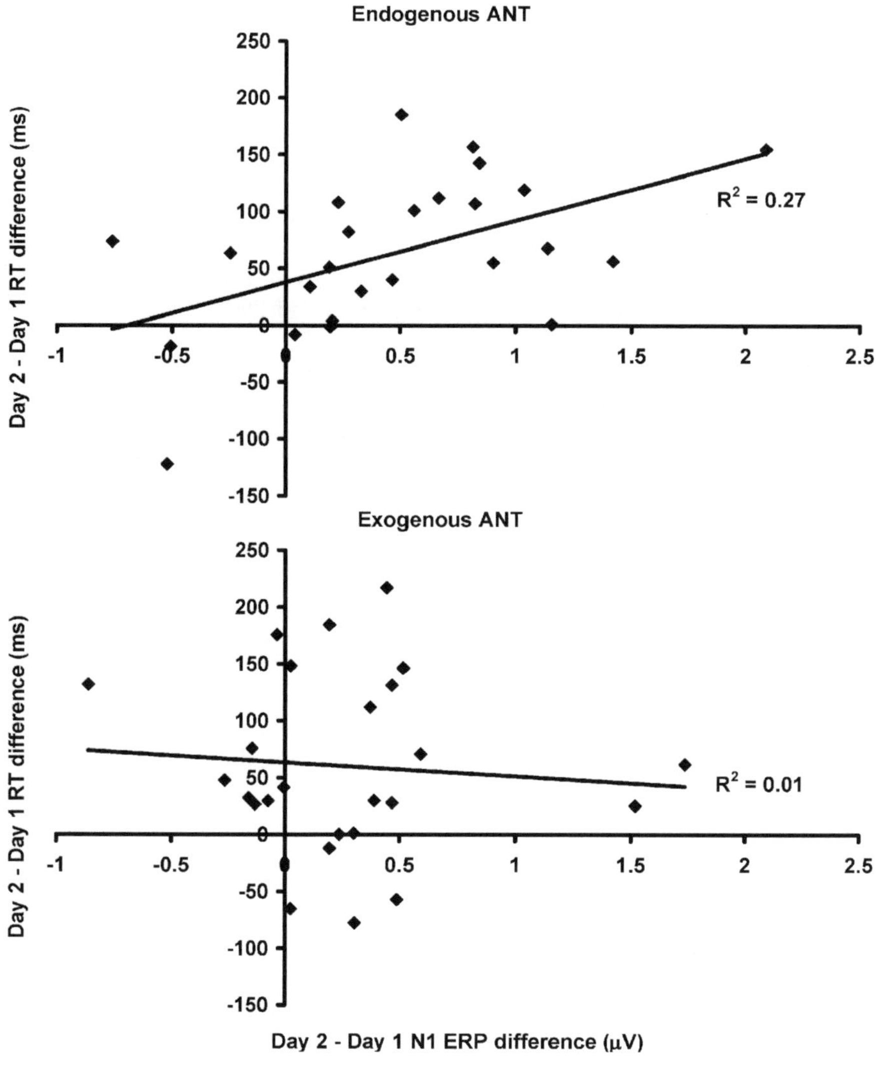

Figure 6.6 Correlation plots between N1 and RT Day 2—Day 1 differences for the Endogenous (top panel) and Exogenous (bottom panel) ANTs

White Matter Differences

The same percent change score approach used in the metacognitive examination above was utilized in examining whether differences in vulnerability were predicted by brain white matter microstructure. Twenty-four West Point Military Academy cadets completed the above-mentioned VMC task both before and after 24 hours of TSD and were split into two groups based on a median split on the percent change in accuracy from pre- and post-TSD on the VMC task was used for analyses (see section above). Participants were divided into those who were more vulnerable (VUL, n = 12) and those who were less vulnerable (NON, n = 12) to TSD.

Whole brain DTI was also acquired for 25 directions using a dual shot echo planar imaging and a twice-refocused spin echo pulse sequence, optimized to minimize eddy current-induced distortions (GE 3T, TR/TE = 12000/71.1, B = 1000, 128 × 128 matrix, 3 mm (0 mm gap) slice thickness, .94 × .94 mm in plane resolution, 1 T2 + 25 DWI). From these images, fractional anisotropy (FA) values were calculated on a voxel-by-voxel basis and then compared between vulnerability groups.

Results from these FA contrasts indicated multiple WM pathways where FA values were significantly greater for the NON group relative to the VUL group. Specifically, higher FA values in the posterior limb and retrolenticular part of the internal capsule, posterior corona radiata, forceps major, posterior thalamic radiation, superior longitudinal fasciculus, and the genu of the corpus callosum (see Figure 6.7) all seem to be associated with decreased vulnerability to TSD. Finally, nearly all regions revealing significant differences also showed significant linear correlations between the percent change score and FA.

Conclusion

The results reviewed here indicate multiple cognitive domains across which important individual differences are revealed in how people respond to conditions of sleep deprivation. The findings of the classification study indicate that those participants who utilize explicit, consciousness-demanding rule-based strategies at the end of Day 1 show a performance deficit during SD while those utilizing the implicit—essentially automatic—procedural system into Day 2 show no performance drop. Metacognitive ability also appears to predict how well an individual will cope with SD as the individuals identified as more vulnerable to SD declined in metacognitive capacity from Day 1 to Day 2 while the individuals less vulnerable to SD did not. In the visual spatial attention experiment, brain bioelectric responses were predictive of an individual's decrements of behavioral performance, as evidenced by increased RTs and errors, after SD. Finally, FA, a measure of WM organization, showed significant correlations in multiple brain regions with an individual's vulnerability to SD, with those participants identified

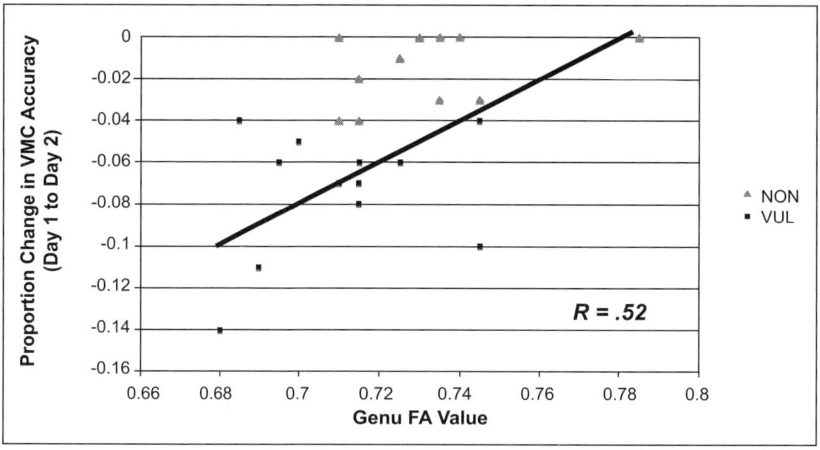

Figure 6.7 Graph showing the correlation between the FA values extracted from the genu of the CC and the proportion change in VMC accuracy between Day 1 and Day 2. Reprinted with permission from the authors

as less vulnerable to SD showing higher FA values (see Rocklage, Williams, Pacheco, and Schnyer, in press, for full results).

There are at least two readily identifiable alternatives in explaining these differences: they may reflect individual differences that are specific to each task or, alternatively, the critical individual difference factor may be a single cognitive component that plays a role across all the domains presented here. In the case of the former, the explanation would be that these tasks show little relation to one another and that the individual differences seen in each task are unrelated. In the case of the latter, it is plausible that there is an underlying cognitive component that is critical in determining how well an individual will respond to SD. Though task differences certainly play some role in the differences seen, given our findings it seems that executive functioning, which has been linked to the prefrontal cortex, may be a good candidate to help explain these differences across tasks.

In the classification study, greater executive control may be required to inhibit inefficient strategies during category learning. Furthermore, individuals who attempt to utilize the more rule-based strategy may fail to perform as well during SD simply because it demands more cognitive control resources. Those more susceptible to SD also showed a decline in metacognitive ability, which is thought to rely on executive functioning (Fernandez-Duque, Baird, and Posner 2000). Prior to separating participants on vulnerability, our findings replicated previous researchers' work showing no metacognitive deficits between days (see Baranski 2007). The decline in accurate reflection was apparent only after taking individual differences in vulnerability to SD into account. The ability to switch attention apart from any external cues may also require executive control as such a task requires a temporal integration of task instructions and visual information. Such integration may be superfluous in the exogenous cue condition where participant attention is automatically captured by a cue appearing at the target location. Moreover, RT increases were significantly correlated with SD-related decreases in the amplitude of the attention-sensitive N1 ERP component in individuals. This finding indicates that an individual's brain bioelectric responses can reflect decrements in behavioral performance due to SD. Finally, though the WM differences seen between vulnerability groups are suggestive of whole brain differences, the genu of the corpus callosum, a WM region connecting the hemispheres of the PFC, was particularly correlated with vulnerability to SD. Thus, though individuals were split into vulnerability groups simply by their ability to remain awake during SD, there seems to be special significance for the frontal area. Executive functioning, and the PFC in particular, then, seem likely candidates for a common factor that accounts for the individual differences seen across these tasks.

There are a number of reasons to believe that executive control may reflect individual differences to a greater degree than other cognitive processes. Research in brain development has indicated that the PFC in particular is more highly influenced by genetics, leading to more potential individual differences in this area than others (Thompson et al. 2001). Moreover, the PFC is one of the last brain structures to develop completely and this may influence the results of studies using primarily college-age participants (Gogtay et al. 2004). As opposed to the occipital lobes, which develop in a relatively linear fashion, grey matter (GM) development in the frontal lobes is nonlinear. Frontal lobe GM volume tends to peak around age 11 or 12 and then declines during adolescence (Gogtay et al. 2004). Meanwhile, there is also an increase in PFC white matter density from puberty and into adulthood (see Paus 2005). It has been theorized that these changes, along with other cellular processes occurring during this time, combine to increase the efficiency of the frontal cognitive system (Blakemore and Choudhury 2006). Considering these vast developmental influences on the PFC, it seems plausible that the functioning of this area would more completely reflect individual differences to SD.

Obtaining a better understanding of the role of cognitive control and executive functioning, mediated at least in part by the PFC, during SD may require a broader range of ages across the developmental trajectory of these functions. In addition, it would be important to implement a collection of additional neuropsychological tests putatively sensitive to executive functioning/PFC differences to be used as task correlates against SD-related performance declines. As demonstrated in these studies, future researchers may be well served to consider individual vulnerability differences to SD in the search for connections between SD, general performance decline, executive functioning/the PFC, and other related phenomena.

References

Ashby, F.G., Alfonso-Reese, L.A., Turken, A.U. and Waldron, E.M. (1998) "A Neuropsychological Theory of Multiple Systems in Category Learning", *Psychol Rev*, 105, 442–81.

Ashby, F.G. and Ennis, J. (2006) "The Role of the Basal Ganglia in Category Learning", *The Psychology of Learning and Motivation*, 471, 1–36.

Ashby, F.G. and Gott, R. (1988) "Decision Rules in the Perception and Categorization of Multidimensional Stimuli", *Journal of Experimental Psychology: Learning, Memory, and Cognition*, 14, 33–53.

Ashby, F.G. and Maddox, W.T. (2005) "Human Category Learning", *Annu Rev Psychol*, 56, 149–78.

Assaf, Y. and Pasternak, O. (2008) "Diffusion Tensor Imaging (DTI)-Based White Matter Mapping in Brain Research: A Review", *J Mol Neurosci*, 34, 51–61.

Baranski, J.V. (2007) "Fatigue, Sleep Loss, and Confidence in Judgment", *J Exp Psychol Appl*, 13, 182–96.

Beaulieu, C. (2002) "The Basis of Anisotropic Water Diffusion in the Nervous System—a Technical Review", *NMR Biomed*, 15, 435–55.

Blakemore, S.J. and Choudhury, S. (2006) "Development of the Adolescent Brain: Implications for Executive Function and Social Cognition", *J Child Psychol Psychiatry*, 47, 296–312.

Chee, M.W. and Choo, W.C. (2004) "Functional Imaging of Working Memory after 24 Hr of Total Sleep Deprivation", *J Neurosci*, 24, 4560–67.

Chee, M.W., Tan, J.C., Zheng, H., Parimal, S., Weissman, D.H., Zagorodnov, V. and Dinges, D.F. (2008) "Lapsing During Sleep Deprivation Is Associated with Distributed Changes in Brain Activation", *J Neurosci*, 28, 5519–28.

Dinges, D.F., Pack, F., Williams, K., Gillen, K.A., Powell, J.W., Ott, G.E., Aptowicz, C. and Pack, A.I. (1997) "Cumulative Sleepiness, Mood Disturbance, and Psychomotor Vigilance Performance Decrements During a Week of Sleep Restricted to 4–5 Hours Per Night", *Sleep*, 20, 267–77.

Fan, J., Byrne, J., Worden, M.S., Guise, K.G., McCandliss, B.D., Fossella, J. and Posner, M.I. (2007) "The Relation of Brain Oscillations to Attentional Networks", *J Neurosci*, 27, 6197–6206.

Fernandez-Duque, D., Baird, J.A. and Posner, M.I. (2000) "Executive Attention and Metacognitive Regulation", *Conscious Cogn*, 9, 288–307.

Filoteo, J.V., Maddox, W.T., Simmons, A.N., Ing, A.D., Cagigas, X.E., Matthews, S. and Paulus, M.P. (2005) "Cortical and Subcortical Brain Regions Involved in Rule-Based Category Learning", *Neuroreport*, 16, 111–15.

Gogtay, N., Giedd, J.N., Lusk, L., Hayashi, K.M., Greenstein, D., Vaituzis, A.C., Nugent, T.F., 3rd, Herman, D.H., Clasen, L.S., Toga, A.W., Rapoport, J.L. and Thompson, P.M. (2004) "Dynamic Mapping of Human Cortical Development During Childhood through Early Adulthood", *Proc Natl Acad Sci USA*, 101, 8174–9.

Gunter, T.C., van der Zande, R.D., Wiethoff, M., Mulder, G. and Mulder, L.J. (1987) "Visual Selective Attention During Meaningful Noise and after Sleep Deprivation", *Electroencephalogr Clin Neurophysiol Suppl*, 40, 99–107.

Heaton, R. (1980) "A Manual for the Wisconsin Card Sorting Test", (Odessa, Florida: Psychological Assessment Resources, Inc.)

Herscovitch, J., Stuss, D. and Broughton, R. (1980) "Changes in Cognitive Processing Following Short-Term Cumulative Partial Sleep Deprivation and Recovery Oversleeping", *Journal of Clinical Neuropsychology*, 2, 301–19.

Jones, D.K., Lythgoe, D., Horsfield, M.A., Simmons, A., Williams, S.C. and Markus, H.S. (1999) "Characterization of White Matter Damage in Ischemic Leukoaraiosis with Diffusion Tensor MRI", *Stroke*, 30, 393–7.

Lim, J. and Dinges, D.F. (2008) "Sleep Deprivation and Vigilant Attention", *Ann N Y Acad Sci*, 1129, 305–22.

Love, B.C. and Gureckis, T.M. (2007) "Models in Search of a Brain", *Cogn Affect Behav Neurosci*, 7, 90–108.

Luck, S.J., Woodman, G.F. and Vogel, E.K. (2000) "Event-Related Potential Studies of Attention", *Trends Cogn Sci*, 4, 432–40.

Madden, D.J., Spaniol, J., Costello, M.C., Bucur, B., White, L.E., Cabeza, R., Davis, S.W., Dennis, N.A., Provenzale, J.M. and Huettel, S.A. (2009) "Cerebral White Matter Integrity Mediates Adult Age Differences in Cognitive Performance", *J Cogn Neurosci*, 21, 289–302.

Maddox, W.T. and Ashby, F.G. (1993) "Comparing Decision Bound and Exemplar Models of Categorization", *Percept Psychophys*, 53, 49–70.

Maddox, W.T. and Ashby, F.G. (2004) "Dissociating Explicit and Procedural-Learning Based Systems of Perceptual Category Learning", *Behav Processes*, 66, 309–32.

Maddox, W.T., Glass, B., Wolosin, S., Savarie, Z., Bowen, C., Matthews, M. and Schnyer, D. (in press) "The Effects of Sleep Deprivation on Information-Integration Categorization Performance", *Sleep*.

Mander, B.A., Reid, K.J., Davuluri, V.K., Small, D.M., Parrish, T.B., Mesulam, M.M., Zee, P.C. and Gitelman, D.R. (2008) "Sleep Deprivation Alters Functioning within the Neural Network Underlying the Covert Orienting of Attention", *Brain Res*, 1217, 148–56.

McCarthy, M.E. and Waters, W.F. (1997) "Decreased Attentional Responsivity During Sleep Deprivation: Orienting Response Latency, Amplitude, and Habituation", *Sleep*, 20, 115–23.

Medina, D., DeToledo-Morrell, L., Urresta, F., Gabrieli, J.D., Moseley, M., Fleischman, D., Bennett, D.A., Leurgans, S., Turner, D.A. and Stebbins, G.T. (2006) "White Matter Changes in Mild Cognitive Impairment and Ad: A Diffusion Tensor Imaging Study", *Neurobiol Aging*, 27, 663–72.

Nomura, E.M., Maddox, W.T., Filoteo, J.V., Ing, A.D., Gitelman, D.R., Parrish, T.B., Mesulam, M.M. and Reber, P.J. (2007) "Neural Correlates of Rule-Based and Information-Integration Visual Category Learning", *Cereb Cortex*, 17, 37–43.

Norton, R. (1970) "The Effects of Acute Sleep Deprivation on Selective Attention", *Br J Psychol*, 61, 157–61.

Paus, T. (2005) "Mapping Brain Maturation and Cognitive Development During Adolescence", *Trends Cogn Sci*, 9, 60–68.

Poldrack, R.A. and Foerde, K. (2008) "Category Learning and the Memory Systems Debate", *Neurosci Biobehav Rev*, 32, 197–205.

Portas, C.M., Rees, G., Howseman, A.M., Josephs, O., Turner, R. and Frith, C.D. (1998) "A Specific Role for the Thalamus in Mediating the Interaction of Attention and Arousal in Humans", *J Neurosci*, 18, 8979–89.

Rocklage, M., Williams, V., Pacheco, J. and Schnyer, D. (in press) "White Matter Differences Predict Cognitive Vulnerability to Sleep Deprivation", *Sleep*.

Schnyer, D.M., Nicholls, L. and Verfaellie, M. (2005) "The Role of VMPC in Metamemorial Judgments of Content Retrievability", *J Cogn Neurosci*, 17, 832–46.

Seger, C.A. (2008) "How Do the Basal Ganglia Contribute to Categorization? Their Roles in Generalization, Response Selection, and Learning Via Feedback", *Neurosci Biobehav Rev*, 32, 265–78.

Seger, C.A. and Cincotta, C.M. (2005) "The Roles of the Caudate Nucleus in Human Classification Learning", *J Neurosci*, 25, 2941–51.

Seger, C.A. and Cincotta, C.M. (2006) "Dynamics of Frontal, Striatal, and Hippocampal Systems During Rule Learning", *Cereb Cortex*, 16, 1546–55.

Thompson, P.M., Cannon, T.D., Narr, K.L., van Erp, T., Poutanen, V.P., Huttunen, M., Lonnqvist, J., Standertskjold-Nordenstam, C.G., Kaprio, J., Khaledy, M., Dail, R., Zoumalan, C.I. and Toga, A.W. (2001) "Genetic Influences on Brain Structure", *Nat Neurosci*, 4, 1253–8.

Tomasi, D., Wang, R.L., Telang, F., Boronikolas, V., Jayne, M.C., Wang, G.J., Fowler, J.S. and Volkow, N.D. (2009) "Impairment of Attentional Networks after 1 Night of Sleep Deprivation", *Cereb Cortex*, 19, 233–40.

Trujillo, L., Kornguth, S. and Schnyer, D.M. (under revision) "An ERP Examination of the Differential Effects of Sleep Deprivation on Exogenously Cued and Endogenously Cued Attention".

Waldron, E.M. and Ashby, F.G. (2001) "The Effects of Concurrent Task Interference on Category Learning: Evidence for Multiple Category Learning Systems", *Psychon Bull Rev*, 8, 168–76.

Walker, M.P. and Stickgold, R. (2004) "Sleep-Dependent Learning and Memory Consolidation", *Neuron*, 44, 121–33.

Wilson, C. (1995) "The Contribution of Cortical Neurons to the Firing Pattern of Striatal Spiny Neurons" in *Models of Information Processing in the Basal Ganglia*, Houk, J., Davis, J. and Beiser, D. (eds), (Cambridge, MA: Bradford).

Zeithamova, D. and Maddox, W.T. (2006) "Dual-Task Interference in Perceptual Category Learning", *Mem Cognit*, 34, 387–98.

Chapter 7

Identification and Prediction of Substantial Differential Vulnerability to the Neurobehavioral Effects of Sleep Loss

David F. Dinges

Namni Goel

For nearly a century, sleep deprivation has posed a logistical challenge and a largely unmitigated risk to maintaining effective performance in sustained military operations. Traditional solutions based on motivation and training, and countermeasures such as caffeine and power naps have helped, but they have not solved the fundamental problem of sleep deprivation's stress or effects or the risks it poses to human performance capability. This is because such solutions have transient benefits and they are not biological substitutes for sleep and they cannot override the escalating sleep debt that accumulates with sustained sleep restriction (Dinges et al. 1997, Van Dongen et al. 2003, and Belenky et al. 2003).

An important new avenue for improving management of sleep deprivation in sustained operations is to identify, and optimally utilize, the recently discovered innate differences among healthy adults in the ability to resist the cognitive effects of sleep loss (Van Dongen et al. 2004a, Viola et al. 2007, and Goel et al. 2009). Seminal experiments we conducted have shown that substantial yet stable differences among individuals in the magnitude of cognitive and behavioral changes that occur with sleep loss have much more to do with differences in the brain's responses to sleep deprivation than with more malleable psychological aspects of behavior such as motivation and training (Van Dongen et al. 2004a, Goel et al. 2009). They also appear to be independent of basal sleep need. While some individuals are highly vulnerable to cognitive performance deficits when sleep deprived, others show remarkable levels of cognitive resistance to sleep loss (Van Dongen et al. 2004b). Identifying the approximate degree of vulnerability/resistance to sleep loss among individuals would permit a greater utilization of personnel and resources, by providing a way of determining those individuals who need countermeasures early and often and those who can withstand longer periods with little to no sleep (Van Dongen et al. 2007). This paper summarizes what is known about these trait-like (phenotypic) individual differences in cognitive vulnerability to sleep deprivation, and efforts to identify objective markers of these differences.

Biological Basis of the Effects of Sleep Loss on Human Performance.

There is extensive evidence that neurobehavioral and cognitive effects of sleep deprivation are in fact mediated by a distributed neurobiological network in the brain involving both the homeostatic drive for sleep and the circadian modulation of wakefulness. Endogenous circadian and homeostatic processes interact to promote periods of stable sleep and wakefulness, with relatively abrupt transitions from one state to the other. Sleep deprivation, which elevates homeostatic sleep drive, leads to a breakdown in this stability, beginning with transient and involuntary intrusions of sleep into periods of wakefulness. These intrusions can be observed through a number of neurobehavioral phenomena, including microsleeps, sleep attacks (involuntary naps), slow eyelid closures, voluntary naps, and slow rolling eye movements (Durmer and Dinges 2005, Goel et al. *in press*). Collectively they reflect the wake state instability that underlies many of the cognitive effects of sleep deprivation (Lim and Dinges 2008).

The stability of wakefulness and the consolidation of sleep are controlled by a set of complex neural pathways. There is currently no singular unifying theory describing their interactions, although their neurobehavioral outputs and effects on neurobehavioral functions are increasingly predictable through mathematical models of these processes (McCauley et al. 2009). The ascending cholinergic reticulothalamocortical pathway, which originates in the upper pons, pedunculopontine, and lateral dorsal tegmental nuclei, and activates the thalamus and cerebral cortex, has a major role in maintaining arousal and wakefulness, although histaminergic, noradrenergic, dopaminergic, and serotonergic mechanisms likely also play a role. Conversely, the ventrolateral preoptic nucleus (VLPO), an area rich in galanin- and GABA-containing neurons, is active during sleep, and is a highly accurate real-time marker of sleep duration. The ascending cholinergic pathway and VLPO are mutually inhibitory, thus forming a "flip-flop" switch that when activated tends to stay active by inhibiting the opposing pathway (Saper et al. 2005). A small perturbation in the system may lead to a sudden change in the pathway whose activity is dominant (for example, instability can occur under certain conditions). There is also evidence that orexin-hypocretin neurons are responsible for both stabilizing the sleep switch and altering its equilibrium point (from promoting wakefulness to promoting sleep and vice versa). These molecules interact directly with the arousal system, but not with the VLPO, suggesting that their action inhibits unwanted lapses into sleep (Mignot 2004). Adenosinergic mechanisms also likely participate in the regulation of sleep homeostasis (Porkka-Heiskanen et al. 2000, Strecker et al. 2000), and the output of the suprachiasmatic nuclei (the endogenous circadian clock) modulates the stability of waking neurobehavioral functions.

The Central Role of Wake-State Instability in the Cognitive Effects of Sleep Deprivation

The effects of sleep loss on cognitive performance are primarily manifested as increasing variability in cognitive speed and accuracy on a wide range of tasks (Durmer and Dinges, 2005, Banks and Dinges 2007, Goel et al. *in press*). Thus, behavioral responses become unpredictable with increasing amounts of fatigue from sleep pressure. When this increased performance variability occurs as a result of sleep deprivation (acute or chronic, partial or total), it reflects wake state instability (Doran et al. 2001). *Wake state instability* refers to moment-to-moment shifts in the relationship between neurobiological systems mediating wake maintenance and those mediating sleep initiations (Lim and Dinges 2008). The increased propensity for sleep as well as the tendency for performance to show behavioral lapses, response slowing, time-on-task decrements, and errors of commission (Doran et al. 2001) are signs that sleep-initiating mechanisms deep in the brain are activating during wakefulness. Thus, the cognitive performance variability that is the hallmark of sleep deprivation (Dinges and Kribbs 1991) appears to reflect state instability (Dorrian, Rogers, and Dinges 2005, Durmer and Dinges 2005, Goel et al. *in press*). Sleep-initiating mechanisms repeatedly interfere with wakefulness, making cognitive performance increasingly variable and dependent on compensatory measures, such as motivation, which cannot override elevated sleep pressure without consequences such as errors of commission, which increase as subjects try to avoid errors of omission—lapses (Durmer and Dinges 2005, Doran et al. 2001, Goel et al. *in press*).

Cognitive Effects are only One Aspect of the Neurobehavioral Effects of Sleep Loss

Intrusions of sleep-initiating biology into the neural mechanisms of goal-directed waking performance are evident as increases in a variety of neurobehavioral phenomena: lapses of attention, sleep attacks, increased frequency of voluntary naps, shortened sleep latency, slow eyelid closures and slow-rolling eye movements, and intrusive daydreaming while engaged in cognitive work (Dinges and Kribbs 1991, Banks and Dinges 2007). These phenomena can occur in healthy sleep-deprived people engaged in potentially dangerous activities (such as sleepiness-related motor vehicle crashes have a fatality rate and injury severity level similar to alcohol-related crashes (Pack et al. 1995), and sleep deprivation produces psychomotor impairments equivalent to those induced by alcohol consumption at or above the legal limit (Durmer and Dinges 2005, Goel et al. *in press*). Both acute total sleep deprivation and chronic partial sleep deprivation degrade many aspects of neurocognitive performance (Dinges and Kribbs 1991, Harrison and Horne 2000, Dorrian et al. 2005, Durmer and Dinges 2005, Banks and Dinges 2007, Goel et al. *in press*).

Psychomotor Vigilance Test (PVT) Performance Reflects Wake State Instability

Neurobehavioral deficits in response to acute total and chronic partial sleep deprivation occur in healthy adults and are especially evident in vigilant attention performance, which has proven to be a particularly useful cognitive assay for tracking and mathematically modeling the dynamic modulation of cognitive function by the interaction of the neurobiology of the homeostatic sleep drive and the circadian pacemaker. The Psychomotor Vigilance Test (PVT) that we developed (Dinges and Powell 1985) has proven to be among the most sensitive assays for detecting wake state instability and the cognitive effects of sleep loss (Dorrian et al. 2005, Lim and Dinges 2008), outperforming many cognitive tests that are confounded by aptitude and learning (Balkin et al. 2004). PVT performance lapses reflect the dynamic interactions of sleep homeostatic drive and circadian neurobiology. Thus, PVT lapses of attention reflect the nonlinear interaction of sleep homeostatic drive and circadian dynamics during total sleep deprivation (Van Dongen and Dinges 2005), and the cumulative increase in homeostatic sleep pressure across days of chronic sleep restriction (Van Dongen et al. 2003, Belenky et al. 2003, Goel et al. 2009). The more PVT lapses that occur, the longer the duration of the lapses; these ultimately progress to full-blown sleep attacks (Lim and Dinges 2008). The occurrence of lapses of attention during performance tasks involves reduced neural activation in a distributed brain network that includes frontal control areas, visual sensory cortex and the thalamus (Chee et al. 2008). The default mode network has also been implicated in mediating lapses (Drummond et al. 2005). The PVT has been used to identify large inter-individual differences in response to sleep deprivation by reflecting proportionality between the mean and variance in PVT reaction times as sleep loss progresses (Dinges and Kribbs 1991, Doran et al. 2001).

Substantial Individual Differences in Neurobehavioral Vulnerability to Sleep Loss

Our laboratory recently demonstrated experimentally that differential cognitive vulnerability to sleep deprivation is not random, but rather is stable and trait-like, strongly suggesting a genetic component (Van Dongen et al. 2004a). The intraclass correlation (ICC) coefficients, which express the proportion of variance in the data that is explained by systematic interindividual variability, revealed that stable (trait-like) responses accounted for 58 percent and 68 percent of the overall variance in PVT lapses between multiple sleep-deprivation exposures in the same subjects (Dijkman et al. 1997, Van Dongen et al. 1999, 2003, 2004b). Thus subjects who had high lapse rates during sleep deprivation after one exposure also had high lapse rates during a second exposure, and similarly, those with low lapse rates during one exposure had low lapse rates during a second exposure. The fact that these high ICCs were found when the subjects were exposed to sleep deprivation two to three times under markedly different conditions, such as high versus low stimulation (Dijkman et al. 1997), or 6–hour versus 12–hour sleep time per night (Van Dongen et al. 2004a), strongly supports the conclusion that the marked differences in cognitive vulnerability to sleep deprivation are phenotypic. A phenotype is an observed trait or characteristic of an organism (for example, morphology, development, and behavior) that includes characteristics made visible by some technical procedure (in this case, by sleep deprivation). Phenotypes are products of variations in genes (genotypes), but are also influenced by extragenetic or environmental factors. Phenotypic variation due to underlying heritable genetic variation is a fundamental prerequisite for evolution by natural selection. Thus, our seminal work in this area established the phenotype for differential cognitive vulnerability to sleep loss and showed that its heritability was as substantial as, but independent of, heritability for sleep duration or sleep timing (Van Dongen, Vitellaro, and Dinges 2005). We have also shown that these phenotypic differences in cognitive responses to sleep deprivation are not reliably accounted for by demographic factors (such as age, sex, IQ), by baseline functioning, by aspects of habitual sleep timing, by circadian phase preference, or by any other investigated factor (Doran et al. 2001, Van Dongen et al. 1999, 2004a).

Other studies have confirmed the presence of large inter-individual differences in how severely sleep deprivation affects cognitive functions in healthy adults (Leproult et al. 2003, Van Dongen, Vitellaro, and Dinges 2005, Van Dongen, Caldwell, and Caldwell 2006, Van Dongen 2006, Bliese, Wesensten, and Balkin 2006, Killgore et al. 2009, Frey, Badia, and Wright 2004). There is good evidence that cognitive tasks sensitive to aptitude and learning closely co-vary in response to sleep loss (Van Dongen et al. 2004a), and that differential vulnerability to the cognitive effects of sleep deprivation is associated with the general tendency of vulnerable subjects to have more wake state instability as measured by PVT lapses of attention (Dorrian et al. 2005).

Such differential vulnerability was also found in major experiments we recently completed on chronic partial sleep deprivation (Goel et al. 2009). It is important to recognize that the stable variance accounted for by individual differences in the magnitude of cognitive changes with sleep deprivation is often considerable and comparable to, or larger than, the effect sizes of many experimental and clinical interventions. Moreover, the differential effects are found even in healthy adults who sleep the same duration each night and otherwise have comparable normal cognitive capability when not sleep deprived (Van Dongen et al. 2004a, Leproult et al. 2003). Finding a biomarker for these large and stable cognitive differences in response to sleep deprivation would be a substantial advance in understanding their possible origins and in harnessing the predictability of them for operational scenarios.

Neuroimaging as a Possible Biomarker for Cognitive Vulnerability to Sleep Deprivation.

Although there are a large number of published neuroimaging (functional magnetic resonance imaging {fMRI}, positron emission tomography {PET}) studies on the effects of sleep deprivation on brain activity in healthy adults, only a few studies have examined whether brain images obtained when people are not sleep deprived could be used to identify differential cognitive vulnerability to subsequent sleep deprivation (Mu et al. 2005, Caldwell et al. 2005, Chee et al. 2006, Chuah et al. 2006). The results of these studies have not been entirely consistent, the sample sizes were small, and when brain activation patterns before sleep deprivation were found to predict the magnitude of the cognitive deficits to sleep loss, the amount of variance accounted for was typically below 36 percent. Neuroimaging as a reliable and valid biomarker for accurately identifying phenotypic vulnerability to sleep deprivation will have to meet a higher scientific evidence bar than is currently provided by these investigations—larger samples, repeated sleep-deprivation exposures for better phenotypic characterization, and use of newer imaging techniques are necessary to have any possibility of achieving this goal. If this is done, the data to date suggest neuroimaging may provide one avenue for developing a biomarker for differential vulnerability to the neurobehavioral effects of sleep loss.

Genetics of Differential Vulnerability to Sleep Loss.

Many aspects of sleep need and timing have clear heritability (Van Dongen et al. 2005, Landolt 2008). The stable, trait-like inter-individual differences observed in response to sleep deprivation strongly suggest an underlying genetic component. Recently, two related publications reported on the role of the variable number tandem repeat (VNTR) polymorphism of the circadian gene *PERIOD3 (PER3)* in response to total sleep deprivation. *PER3* is characterized by a 54-nucleotide coding region motif repeating in four or five units and shows similar allelic frequencies in African Americans and Caucasians/European Americans. Compared with the 4-repeat allele ($PER3^{4/4}$; n = 14), the longer, 5-repeat allele ($PER3^{5/5}$; n = 10) was associated with worse cognitive performance during sleep loss and with higher sleep propensity including slow-wave activity (SWA) in the sleep electroencephalogram (EEG)—a putative marker of sleep homeostasis—before and after total sleep deprivation (Viola et al. 2007). A subsequent report on the same subjects clarified that the $PER3^{5/5}$ performance deficits were found only on specific executive function tests, and only at 2–4 hours following the melatonin circadian rhythm peak, from approximately 6–8 AM (Groeger et al. 2008). Such performance differences were hypothesized to be mediated by sleep homeostasis (Viola et al. 2007, Groeger et al. 2008). A recent neuroimaging investigation identified differential brain activation profiles to total sleep deprivation as a function of this *PER3* genotype (Vandewalle et al. 2009).

Although the work on the *PER3* VNTR variants and cognitive responses to total sleep deprivation is important and promising, replication of such data are necessary to consider this polymorphism as a biomarker of cognitive vulnerability to sleep loss. Below we describe evidence that the aforementioned findings may not generalize to responses to chronic partial sleep deprivation (Goel et al. 2009). It is also unlikely that a single genetic variation accounts for all cognitive and homeostatic differences in response to total sleep deprivation (Landolt 2008). For example, one study reported an association between an A2A receptor gene polymorphism and objective and subjective differences in caffeine's effects on non-rapid eye movement (non-REM) sleep after total sleep deprivation (Retey et al. 2007).

Chronic partial sleep deprivation (which is far more common in the general population than total sleep deprivation) also involves differential cognitive vulnerability (Van Dongen et al. 2003, Bliese, Wesensten, and Balkin 2006) and therefore may involve genetic variation. A recent study by our laboratory on the cognitive effects of chronic partial sleep deprivation found no relationship to the *PER3* VNTR genotypes (Goel et al. 2009). That is, $PER3^{4/4}$, $PER3^{4/5}$ and $PER3^{5/5}$ genotypes demonstrated comparable cumulative increases in sleepiness and cumulative decreases in cognitive performance and physiological alertness, across five nights of chronic sleep restriction to four hours per night, with all genotypes showing increasing daily inter-subject variability (Goel et al. 2009). During chronic sleep restriction, $PER3^{5/5}$ subjects exhibited slightly but reliably higher slow-wave energy (SWE) than $PER3^{4/4}$ subjects. In contrast to published data in total sleep deprivation paradigms, the *PER3* VNTR variants did not differ on baseline sleep measures or in their physiological sleepiness, cognitive, executive functioning or subjective responses to chronic sleep restriction. Thus, the *PER3* VNTR polymorphism is not a genetic marker of differential vulnerability to the cumulative neurobehavioral effects of chronic sleep restriction. It is likely that other genetic polymorphisms regulate neurobehavioral responses to chronic sleep restriction.

In conclusion, in recent years—since we originally identified the stable, trait-like (phenotypic) neurobehavioral vulnerability to sleep loss—there has been a growing search for a biomarker of this cognitive vulnerability, in the hope of harnessing this large and critical source of variance in human neurobehavioral responses to sleep deprivation. Recent discoveries in the fields of neuroimaging and genomics suggest the goal is clearly achievable and worth pursuing.

References

Balkin, T.J., Bliese, P.D., Belenky, G., Sing, H., Thorne, D.R., Thomas, M., Redmond, D.P., Russo, M. and Wesensten, N.J. (2004) "Comparative Utility of Instruments for Monitoring Sleepiness-Related Performance Decrements in the Operational Environment", *J Sleep Res*, 13, 219–27.

Banks, S. and Dinges, D.F. (2007) "Behavioral and Physiological Consequences of Sleep Restriction", *J Clin Sleep Med*, 3, 519–28.

Belenky, G., Wesensten, N.J., Thorne, D.R., Thomas, M.L., Sing, H.C., Redmond, D.P., Russo, M.B. and Balkin, T.J. (2003) "Patterns of Performance Degradation and Restoration During Sleep Restriction and Subsequent Recovery: A Sleep Dose-Response Study", *J Sleep Res*, 12, 1–12.

Bliese, P.D., Wesensten, N.J. and Balkin, T.J. (2006) "Age and Individual Variability in Performance During Sleep Restriction", *J Sleep Res*, 15, 376–85.

Caldwell, J.A., Mu, Q., Smith, J.K., Mishory, A., Caldwell, J.L., Peters, G., Brown, D.L. and George, M.S. (2005) "Are Individual Differences in Fatigue Vulnerability Related to Baseline Differences in Cortical Activation?" *Behav Neurosci*, 119, 694–707.

Chee, M.W., Chuah, L.Y., Venkatraman, V., Chan, W.Y., Philip, P. and Dinges, D.F. (2006) "Functional Imaging of Working Memory Following Normal Sleep and after 24 and 35 H of Sleep Deprivation: Correlations of Fronto-Parietal Activation with Performance", *Neuroimage*, 31, 419–28.

Chee, M.W., Tan, J.C., Zheng, H., Parimal, S., Weissman, D.H., Zagorodnov, V. and Dinges, D.F. (2008) "Lapsing During Sleep Deprivation is Associated with Distributed Changes in Brain Activation", *J Neurosci*, 28, 5519–28.

Chuah, Y.M., Venkatraman, V., Dinges, D.F. and Chee, M.W. (2006) "The Neural Basis of Interindividual Variability in Inhibitory Efficiency after Sleep Deprivation", *J Neurosci*, 26, 7156–62.

Dijkman, M., Sachs, N., Levine, E., Mallis, M., Carlin, M., Gillen, K., Powell, J., Samuel, S., Mullington, J., Rosekind, M. and Dinges, D. (1997) "Effects of Reduced Stimulation on Neurobehavioral Alertness Depend on Circadian Phase During Human Sleep Deprivation", *Sleep Research*, 26, 265.

Dinges, D.F. and Kribbs, N.B. (1991) "Performing While Sleepy: Effects of Experimentally-Induced Sleepiness" in *Sleep, Sleepiness, and Performance*, Monk, T.H. (eds), (Chester, England: Wiley), 97–128.

Dinges, D.F., Pack, F., Williams, K., Gillen, K.A., Powell, J.W., Ott, G.E., Aptowicz, C. and Pack, A.I. (1997) "Cumulative Sleepiness, Mood Disturbance, and Psychomotor Vigilance Performance Decrements During a Week of Sleep Restricted to 4–5 Hours Per Night", *Sleep*, 20, 267–77.

Dinges, D.F. and Powell, J.W. (1985) "Microcomputer Analyses of Performance on a Portable, Simple Visual RT Task During Sustained Operations", *Behavior Research Methods, Instruments, & Computers*, 17, 652–5.

Doran, S.M., Van Dongen, H.P. and Dinges, D.F. (2001) "Sustained Attention Performance During Sleep Deprivation: Evidence of State Instability", *Arch Ital Biol*, 139, 253–67.

Dorrian, J., Rogers, N.L. and Dinges, D.F. (2005) "Psychomotor Vigilance Performance: A Neurocognitive Assay Sensitive to Sleep Loss" in *Sleep Deprivation: Clinical Issues, Pharmacology and Sleep Loss Effects*, Kushida, C. (eds), (New York, NY: Marcel Dekker, Inc.), 39–70.

Drummond, S.P., Bischoff-Grethe, A., Dinges, D.F., Ayalon, L., Mednick, S.C. and Meloy, M.J. (2005) "The Neural Basis of the Psychomotor Vigilance Task", *Sleep*, 28, 1059–68.

Durmer, J.S. and Dinges, D.F. (2005) "Neurocognitive Consequences of Sleep Deprivation", *Semin Neurol*, 25, 117–29.

Frey, D.J., Badia, P. and Wright, K.P., Jr. (2004) "Inter- and Intra-Individual Variability in Performance near the Circadian Nadir During Sleep Deprivation", *J Sleep Res*, 13, 305–15.

Goel, N., Banks, S., Mignot, E. and Dinges, D.F. (2009) "Per3 Polymorphism Predicts Cumulative Sleep Homeostatic but Not Neurobehavioral Changes to Chronic Partial Sleep Deprivation", *PLoS One*, 4, e5874.

Goel, N., Rao, H., Durmer, J.S. and Dinges, D.F. (*in press*) "Neurocognitive Consequences of Sleep Deprivation", *Seminars in Neurology*, 29.

Groeger, J.A., Viola, A.U., Lo, J.C., von Schantz, M., Archer, S.N. and Dijk, D.J. (2008) "Early Morning Executive Functioning During Sleep Deprivation is Compromised by a Period3 Polymorphism", *Sleep*, 31, 1159–67.

Harrison, Y. and Horne, J.A. (2000) "The Impact of Sleep Deprivation on Decision Making: A Review", *J Exp Psychol Appl*, 6, 236–49.

Killgore, W.D., Grugle, N.L., Reichardt, R.M., Killgore, D.B. and Balkin, T.J. (2009) "Executive Functions and the Ability to Sustain Vigilance During Sleep Loss", *Aviat Space Environ Med*, 80, 81–7.

Landolt, H.P. (2008) "Genotype-Dependent Differences in Sleep, Vigilance, and Response to Stimulants", *Curr Pharm Des*, 14, 3396–3407.

Leproult, R., Colecchia, E.F., Berardi, A.M., Stickgold, R., Kosslyn, S.M. and Van Cauter, E. (2003) "Individual Differences in Subjective and Objective Alertness During Sleep Deprivation Are Stable and Unrelated", *Am J Physiol Regul Integr Comp Physiol*, 284, R280–290.

Lim, J. and Dinges, D.F. (2008) "Sleep Deprivation and Vigilant Attention", *Ann NY Acad Sci*, 1129, 305–22.

McCauley, P., Kalachev, L.V., Smith, A.D., Belenky, G., Dinges, D.F. and Van Dongen, H.P. (2009) "A New Mathematical Model for the Homeostatic Effects of Sleep Loss on Neurobehavioral Performance", *J Theor Biol*, 256, 227–39.

Mignot, E. (2004) "Sleep, Sleep Disorders and Hypocretin (Orexin)", *Sleep Med*, 5 Suppl 1, S2–8.

Mu, Q., Mishory, A., Johnson, K.A., Nahas, Z., Kozel, F.A., Yamanaka, K., Bohning, D.E. and George, M.S. (2005) "Decreased Brain Activation During a Working Memory Task at Rested Baseline is Associated with Vulnerability to Sleep Deprivation", *Sleep*, 28, 433–46.

Pack, A.I., Pack, A.M., Rodgman, E., Cucchiara, A., Dinges, D.F. and Schwab, C.W. (1995) "Characteristics of Crashes Attributed to the Driver Having Fallen Asleep", *Accid Anal Prev*, 27, 769–75.

Porkka-Heiskanen, T., Strecker, R.E. and McCarley, R.W. (2000) "Brain Site-Specificity of Extracellular Adenosine Concentration Changes During Sleep Deprivation and Spontaneous Sleep: An *in vivo* Microdialysis Study", *Neuroscience*, 99, 507–17.

Retey, J.V., Adam, M., Khatami, R., Luhmann, U.F., Jung, H.H., Berger, W. and Landolt, H.P. (2007) "A Genetic Variation in the Adenosine A2A Receptor Gene (Adora2a) Contributes to Individual Sensitivity to Caffeine Effects on Sleep", *Clin Pharmacol Ther*, 81, 692–8.

Saper, C.B., Scammell, T.E. and Lu, J. (2005) "Hypothalamic Regulation of Sleep and Circadian Rhythms", *Nature*, 437, 1257–63.

Strecker, R.E., Moriarty, S., Thakkar, M.M., Porkka-Heiskanen, T., Basheer, R., Dauphin, L.J., Rainnie, D.G., Portas, C.M., Greene, R.W. and McCarley, R.W. (2000) "Adenosinergic Modulation of Basal Forebrain and Preoptic/Anterior Hypothalamic Neuronal Activity in the Control of Behavioral State", *Behav Brain Res*, 115, 183–204.

Van Dongen, H.P. (2006) "Shift Work and Inter-Individual Differences in Sleep and Sleepiness", *Chronobiol Int*, 23, 1139–47.

Van Dongen, H.P.A. and Dinges, D.F. (2005) "Circadian Rhythm in Sleepiness, Alterness, and Performance" in *Principles and Practice of Sleep Medicine, 4th Ed.*, Kryger, M.H., Roth, T. and Dement, W.C. (eds), (Philadelphia, PA: W.B. Saunders), 435–43.

Van Dongen, H.P., Baynard, M.D., Maislin, G. and Dinges, D.F. (2004a) "Systematic Interindividual Differences in Neurobehavioral Impairment from Sleep Loss: Evidence of Trait-Like Differential Vulnerability", *Sleep*, 27, 423–33.

Van Dongen, H.P., Caldwell, J.A., Jr. and Caldwell, J.L. (2006) "Investigating Systematic Individual Differences in Sleep-Deprived Performance on a High-Fidelity Flight Simulator", *Behav Res Methods*, 38, 333–43.

Van Dongen, H.P., Maislin, G. and Dinges, D.F. (2004b) "Dealing with Inter-Individual Differences in the Temporal Dynamics of Fatigue and Performance: Importance and Techniques", *Aviat Space Environ Med*, 75, A147–154.

Van Dongen, H.P., Maislin, G., Mullington, J.M. and Dinges, D.F. (2003) "The Cumulative Cost of Additional Wakefulness: Dose-Response Effects on Neurobehavioral Functions and Sleep Physiology from Chronic Sleep Restriction and Total Sleep Deprivation", *Sleep*, 26, 117–26.

Van Dongen, H.P., Mott, C.G., Huang, J.K., Mollicone, D.J., McKenzie, F.D. and Dinges, D.F. (2007) "Optimization of Biomathematical Model Predictions for Cognitive Performance Impairment in Individuals: Accounting for Unknown Traits and Uncertain States in Homeostatic and Circadian Processes", *Sleep*, 30, 1129–43.

Van Dongen, H.P., Vitellaro, K.M. and Dinges, D.F. (2005) "Individual Differences in Adult Human Sleep and Wakefulness: Leitmotif for a Research Agenda", *Sleep*, 28, 479–96.

Van Dongen, H.P.A., Dijkman, M.V., Maislin, G. and Dinges, D.F. (1999) "Phenotypic Aspect of Vigilance Decrement During Sleep Deprivation", *Physiologist*, 42, A–5.

Vandewalle, G., Archer, S.N., Wuillaume, C., Balteau, E., Degueldre, C., Luxen, A., Maquet, P. and Dijk, D.J. (2009) "Functional Magnetic Resonance Imaging-Assessed Brain Responses During an Executive Task Depend on Interaction of Sleep Homeostasis, Circadian Phase, and PER3 Genotype", *J Neurosci*, 29, 7948–56.

Viola, A.U., Archer, S.N., James, L.M., Groeger, J.A., Lo, J.C., Skene, D.J., von Schantz, M. and Dijk, D.J. (2007) "PER3 Polymorphism Predicts Sleep Structure and Waking Performance", *Curr Biol*, 17, 613–18.

Chapter 8
Sustaining Performance: The Other Side of Sleep

Robert Stickgold

Introduction

Sustaining performance across extended periods of time requires the maintenance of both physical and mental efficiency in the face of conditions that would normally lead to their impairment. While the importance of maintaining cognitive functioning in the face of stressors such as sleep deprivation is well recognized, the importance of sustaining memory processes is relatively unappreciated. But the ability to remember new facts, learn new skills and understand a rapidly changing situation is equally important. In many circumstances, sustained performance requires rapid changes is how one acts, based on a rapidly changing situation. Individuals need to be able to learn rapidly in a changing environment, and must also be able to understand how to optimally use this new information. Sleep is critical to both of these functions.

A wide range of studies has led to a growing appreciation of the role of sleep in learning and memory consolidation. Sleep appears to act both prospectively and retrospectively in facilitating memory processing. It both prepares the brain for subsequent memory encoding and subsequently facilitates the consolidation of recently learned information and skills. In what follows, we review work, primarily carried out in my laboratory, which provide evidence for this claim.

Procedural Skill Learning

The most robust evidence of sleep-dependent memory consolidation in humans comes from studies of procedural skill learning. The hallmark of procedural learning is that it occurs without the critical participation of the hippocampal formation, and develops over repeated trials of the task. Sleep-dependent procedural skill learning has been documented in the perceptual, motor, and cognitive domains. Here we review data for perceptual and motor skill learning, returning to cognitive procedural learning later.

The Visual Texture Discrimination Task (TDT)

The first evidence of sleep-dependent consolidation of perceptual skill learning involved the texture discrimination task of Karni and Sagi (1991). It requires subjects to identify the orientation of a 3-element array of diagonal bars against a background of horizontal bars (Figure 8.1). Karni et al. (1994) reported an improvement on the TDT that appeared to occur across periods of both wake and sleep. But the overnight improvement failed to occur if rapid-eye-movement sleep (REM) was disrupted, leading them to conclude that REM sleep, but not slow wave sleep (SWS), supported this slow consolidation of TDT learning.

In more extensive studies (Stickgold, James, and Hobson 2000, Stickgold et al. 2000), improvement was demonstrated to occur only across periods of sleep, and not over wake (Figure 8.2A). In addition, overnight improvement was strongly correlated with the product of the amounts of slow wave sleep (SWS) obtained early in the night and REM sleep obtained late in the night (r=0.89; p<0.0001; Figure 8.2B). Thus, improvement appeared to require a full eight-hour night of sleep, with SWS during the first quarter of the night and REM during the last quarter both being necessary. Indeed, subjects reporting less than six hours of sleep showed no overnight improvement (Stickgold et al. 2000), suggesting that a minimum of six hours of sleep was required to generate such improvement.

But subsequent nap studies belied this conclusion. Subjects who succeeded in obtaining both SWS and REM in a 60 or 90 minutes afternoon nap opportunity showed as much improvement when tested several hours later as did subjects allowed a full night of sleep (Mednick, Nakayama, and Stickgold 2003). Thus, afternoon naps seem to provide more efficient consolidation of visual procedural learning than does nocturnal sleep.

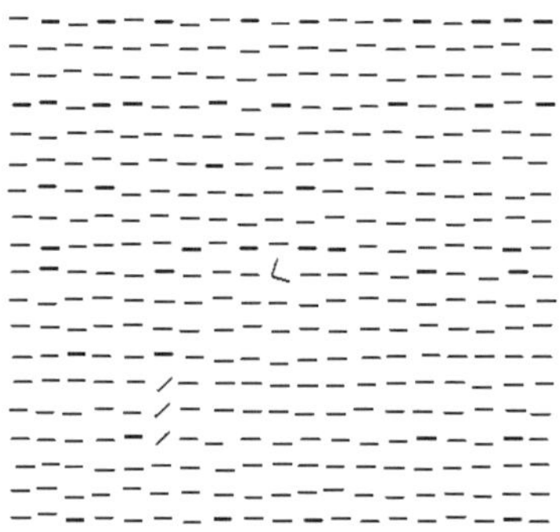

Figure 8.1 The visual texture discrimination task (TDT)

A pattern of horizontal bars with an embedded array of 3-diagonal bars is displayed for 16 ms, followed by a blank screen for a variable interstimulus interval, before a mask screen is displayed for 50 ms. In half the trials, the diagonal bar array is vertically oriented in a column (as in this example), and in half horizontally in a row.

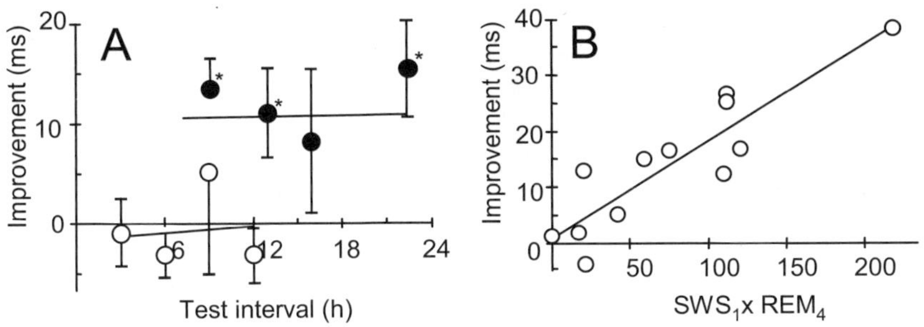

Figure 8.2 Sleep-dependent TDT improvement

(A) Improvement on the TDT was only seen after a night of sleep (filled circles) and not after equivalent periods of daytime wake (open circles); (B) Overnight improvement correlated strongly (r =.89) with the product of SWS in the first quarter and REM in the last quarter of an eight hour night. *p <.05. (Stickgold et al., 2000)

The Finger-Tapping Motor Sequence Task (MST)

Evidence for similar sleep-dependent improvement on a motor task was first reported on a sequential finger-tapping task. As with the TDT, time-dependent improvement on this motor sequence task (MST) showed an absolute requirement for sleep (Walker et al. 2002, 2003).

In the MST, subjects repeatedly type a sequence of digits (for example, the 5-digit sequence 4-1-3-2-4) with their non-dominant, left hand, on a computer keyboard, as quickly and accurately as possible, performing twelve 30-sec trials separated by 30-sec rest periods. Performance improves rapidly over the first three or four trials, but increases little thereafter. At test, subjects perform two or three trials, and the improvement from the end of training to retest is calculated. Plateau levels are similar whether subjects are trained in the morning or evening (Figure 8.3A, left-hand bar in each panel). When tested 12 hours later, only the group trained in the evening and tested after a night of sleep (Fig 3A, left panel) showed improvement. However, when retested after an additional 12 hours, those subjects initially trained in the morning, and now retested after a night of sleep, showed significant improvement (Figure 8.3A, right panel). Thus, regardless of time of training, improvement was only observed after sleep. When subjects trained in the evening had their sleep recorded overnight, their subsequent improvement was strongly correlated with the amount of Stage 2 sleep during the intervening night, primarily in the last quarter of the night (r = 0.72, p = 0.0008; Figure 8.3B).

As with the visual TDT, improvement can be seen more rapidly with daytime naps (Nishida and Walker 2007). Here again, significant improvement was seen after 60–90 minutes of sleep, equivalent to that normally seen across an entire night of sleep.

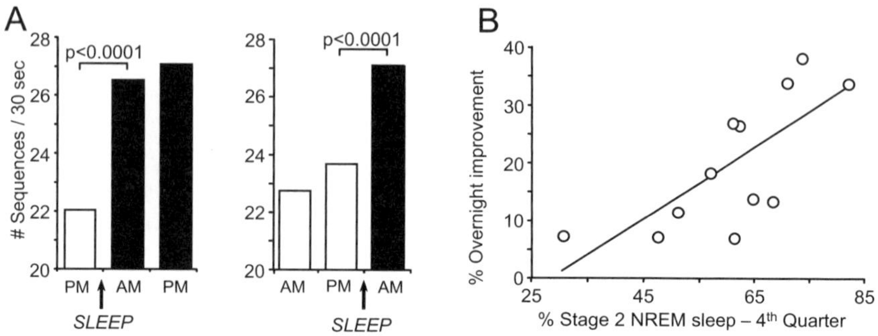

Figure 8.3 Sleep-dependent MST improvement

(A) Improvement on the MST was only seen after a night of sleep (black bars) and not after equivalent periods of daytime wake (white bars); (B) Overnight improvement correlated strongly ($r = 0.79$) with the amount of Stage 2 NREM sleep in the last quarter of an eight hour night. (Walker et al. 2002)

Interim Summary

For both perceptual and motor procedural skills, sleep leads to an absolute improvement in performance, with performance at the start of post-sleep testing superior to that seen at the end of training. In both cases, this is a true improvement in ability, and does not simply reflect better performance after rest, as equivalent periods of daytime wake do not yield any significant improvement. Sleep-dependent improvement is seen not only for motor sequence tasks, but for motor adaptation as well (Huber et al. 2004). Interestingly in this case, improvement was correlated with the amount of slow wave activity (EEG spectral power in the 1–4 Hz frequency band) in the first quarter of the night, but not with the amount of time spent in SWS. Sleep-dependent improvement has also been seen for auditory learning (Atienza, Cantero, and Stickgold 2004, Gaab et al. 2004), other sensory modalities have not been tested.

While data from overnight studies of both visual and motor skill learning suggest that a full night of sleep is needed for optimal improvement, in both cases 60–90 minute naps have been shown to produce equivalent improvement. Thus, improvements in performance, which in some circumstances could be a crucial component of sustained performance, can be achieved under some circumstances with surprisingly little sleep.

Emotional Declarative Memory

Declarative memories also benefit from sleep. Deterioration over time is generally greater across periods of wake than across periods including sleep, and appears to depend on periods of SWS (Gais and Born 2004, Plihal and Born 1997, Wagner, Gais, and Born 2001), and sleep appears to make such memories more resistant to subsequent interference (Ellenbogen et al. 2009). As with procedural learning, declarative memories can benefit from naps (Tucker, 2006, Ellenbogen et al. 2006), with as little as ten minutes of sleep producing memory benefits (Brooks and Lack 2006).

Emotional declarative memory appears to represent a special case, depending on REM sleep, rather than SWS (Wagner, Gais, and Born 2001). In this case, studies have shown impacts both for sleep obtained prior to training and for sleep after training.

Sleep Deprivation and Emotional Memory Bias

As might be expected, sleep deprivation prior to training on a declarative memory task leads to impaired retention of memories. But what is perhaps surprising is that the impairment is not equal across emotional valences. Instead, sleep deprivation leads to a selective retention of subsequently encoded negative emotional memories, over both neutral and positive memories (Walker and Stickgold 2006, Yoo et al. 2007). Subjects viewed a series of emotionally positive, neutral, and negative pictures, rating their emotionality, and only incidentally encoding memories of the pictures. Half the subjects had been sleep deprived the night prior to training, and all were allowed two nights of recovery sleep before testing for recognition memory of the viewed slides (Figure 8.4A).

At test, control subjects (Figure 8.4B, solid bars), who had slept normally the night prior to training, performed 76 percent better at recognizing the emotional pictures than the neutral pictures, with no significant difference between the positive and negative images. Subjects who were sleep deprived prior to encoding showed impaired recognition memory at retest, performing 31 percent worse overall than controls. But most striking was the shift in relative recognition for different types of pictures (Figure 8.4B, open bars). While recognition of positive and neutral pictures dropped to similar levels, recognition of negative pictures remained relatively intact, being on average, 104 percent better than for neutral and positive memories.

Emotional Trade-off Task

The sleep deprivation study described above points to a new and important concept in understanding the nature of sleep-dependent memory processing. While a night of sleep deprivation caused an overall 31 percent decrease in subsequent memory encoding, it also produced a qualitative shift in the memory, in this case creating a bias toward recognition of negative memories. This is an aspect of sleep-dependent memory processing that we shall continue to see later. Rather than simply enhancing recall (or impairing encoding), sleep produces qualitative changes in memories. This can be seen in post-encoding, sleep-dependent consolidation of emotional declarative memories.

When subjects view emotional and neutral scenes, all of which have neutral backgrounds but half of which have aversive objects in the foreground, an emotional memory trade-off results. Shortly after viewing the scenes, subjects show enhanced recognition memory for the emotional foreground objects they viewed, but impaired recognition of the neutral backgrounds on which they were displayed, compared to their recognition of emotionally neutral foreground objects and their neutral backgrounds (Kensinger, Garoff-Eaton, and Schacter 2007).

When recognition is tested not 30 minutes after training, but 12 hours later, across wake or sleep, recognition of both foreground objects and backgrounds showed deterioration of approximately 8–10 percent, with the exception of the recognition of the aversive emotional objects after sleep. In contrast to all other values, recognition for these objects improves slightly after sleep, and is significantly better than for objects and backgrounds in all other conditions (Figure 8.5A).

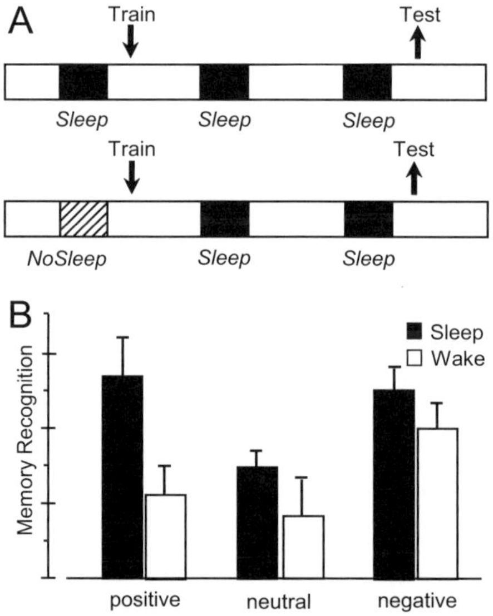

Figure 8.4 Emotional memory bias after sleep deprivation

(A) subjects were trained and then retested 48 hours later, after two nights of normal sleep. Control subjects also had normal sleep the night prior to encoding, while the experimental group was sleep deprived, and trained after an average of 35 hours without sleep; (B) scores for recognition of positive, neutral, and negative memories were biased toward the negative in the sleep deprived group, even 48 hours after encoding (Walker and Stickgold 2006, reprinted with permission, from the Annual Review of Psychology, Volume 57 © 2006 by Annual Reviews).

Interim Summary

Emotional memory is modulated by the extent of sleep both before and after encoding. After a night of sleep deprivation, subjects encode waking events with a strong emotional bias toward the negative. At retest 48 hours later, after two nights of normal recovery sleep, subjects recognized twice as many negative pictures as positive ones, even though they saw equal numbers at encoding. This is a striking and important difference. While not investigated, such a dramatic shift in emotional bias presumably would impair an individual's ability to make appropriate decisions based on these prior experiences, an ability that would be crucial if optimal performance is to be sustained.

Similarly, sleep after encoding appears to block the otherwise normal time-dependent deterioration of the emotional core of complex scenes. As with sleep deprivation prior to encoding, sleep after encoding produces a qualitative change in emotional memories, effectively unbinding the episodic memory for the scene, and selectively retaining its emotional core. This could very well represent the first step in a complex series of changes that episodic memories must undergo to become semanticized, producing a more general semantic memory that would be of greater future value than a highly detailed memory of a specific scene.

Extraction of Gist and Insight

Absolute improvement in performance after sleep, such as that clearly seen with procedural learning, remains more the exception than the rule. Tests of sleep-dependent consolidation of declarative memory rarely show such a pattern. Rather, sleep seems to result in a decreased deterioration over time. But even here, exceptions can be found. Emotional declarative memories seem to be maintained at levels not different from their pre-sleep values after a night of sleep, while recognition of other aspects of the memories deteriorate.

In examples of complex cognitive procedural learning, similar maintenance or enhancement of initial performance can be seen. These are tasks that require the extraction of gist or discovery of insight for their successful performance.

Gist Extraction in a False Memory Task

Sleep's ability to extract gist for more complex stimuli is seen in a study using the Deese, Roediger, and McDermott false memory paradigm (Deese 1959, Roediger and McDermott 1995). Subjects listened to eight lists of words, with instructions to remember the words for later testing. Each list consisted of 12 words strongly related to a single central word, which was not itself included in the list. Thus, they heard 96 words that were strong associates of eight additional words that represented the gist of the overall list. At retest, subjects were given a blank sheet of lined paper and instructed to write down as many words as possible from the lists they heard previously. Retesting occurred either 20 minutes or 12 hours after hearing the lists.

After just 20 minutes, subjects recalled 25 percent of the 96 studied words and falsely recalled 44 percent of the eight gist words (Payne et al. 2006). When recall after 12 hours of daytime wake or across a night of normal sleep was compared to these rates of recall after 20 minutes, a pattern similar to that found for emotional scenes was observed (Figure 8.5B). Recall of studied words was reduced after either wake or sleep, but significantly more forgetting occurred across the day (Figure 8.5B, left), suggesting that sleep protected these memories against deterioration over time. Recall of the gist words also decreased significantly from baseline, but only across the day (Figure 8.5B, right, open bar). In contrast, recall of gist words actually increased, albeit non-significantly, after sleep, and showed significantly better recall than after wake (Figure 8.5B, right, solid bar).

Interim Summary

Thus, sleep appears to lead to a selective maintenance, or even slight improvement, in memory for material representing the gist of the studied material (Payne et al. 2006). The similarity in memory profiles for this gist memory task and the emotional scenes task (*cf.* Figure 8.5A and 8.5B) is noteworthy. In both cases, sleep acts to maintain, or even slightly enhance, memories for what are arguably the most central and salient aspects of the studied material. The fact that this includes false memories for items not actually encountered in the study phase suggests that sleep-dependent memory processing can result in memories that are less accurate, but still more useful, than those of the night before.

Conclusion

The ability to maintain stable and consistent performance over extended periods of time is of critical importance to sustained performance. But the ability to remember new facts, learn new skills, and understand a rapidly changing situation is equally critical. In many circumstances, sustained performance requires rapid change. Thus, research paradigms that use tasks with no learning curve or which are brought to asymptotic performance early in training tell only half the story. Sleep is known to be critical for the maintenance of stable and consistent performance. But it is also crucial for remembering new facts, learning new skills, and understanding rapidly changing situations.

Sleep is a period of time during which the brain performs extensive analysis and modification of recently formed memories. This includes the enhancement of perceptual and motor skills, even after practice-dependent improvement has reached a plateau. More subtly, it can lead to the selective retention and balancing of emotional memories, and the extraction of the gist of a complex stimulus set.

Appreciating how dramatic these sleep effects are, raises new and important questions about how to measure both sustained performance and the effectiveness of countermeasures developed to permit sustained performance under stress.

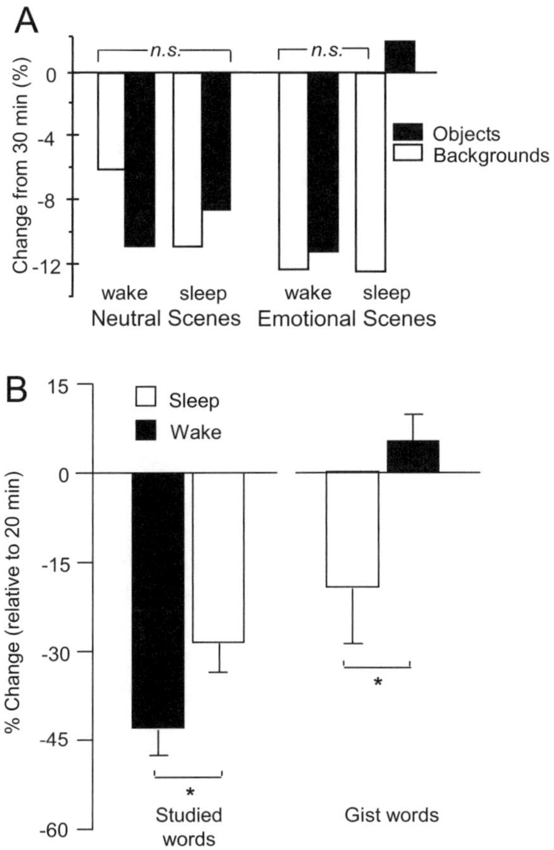

Figure 8.5 Qualitative changes in memory after sleep

(A) Emotional trade-off and sleep: Change in recognition scores 12 hours post-encoding relative to recognition 30 minutes after encoding (separate group; data not shown). Foreground objects (neutral or aversive) were tested separately from their backgrounds (all neutral). Both neutral and emotional scene components deteriorated across wake or sleep, with the exception of emotional objects, which showed a nonsignificant increase after sleep, and significantly better recognition than all other bars. (B) Selective retention of "gist" memory. As with emotional objects in the emotional trade-off paradigm, subjects selectively maintained memory for gist words, representing the theme of each list, preferentially across sleep. (Payne et al. 2009)

References

Atienza, M., Cantero, J.L. and Stickgold, R. (2004) "Posttraining Sleep Enhances Automaticity in Perceptual Discrimination", *J Cogn Neurosci*, 16, 53–64.

Brooks, A. and Lack, L. (2006) "A Brief Afternoon Nap Following Nocturnal Sleep Restriction: Which Nap Duration Is Most Recuperative?", *Sleep*, 29, 831–40.

Deese, J. (1959) "On the Prediction of Occurrence of Particular Verbal Intrusions in Immediate Recall", *J Exp Psychol*, 58, 17–22.

Ellenbogen, J.M., Hulbert, J.C., Jiang, Y. and Stickgold, R. (2009) "The Sleeping Brain's Influence on Verbal Memory: Boosting Resistance to Interference", *PLoS ONE*, 4, e4117.

Ellenbogen, J.M., Hulbert, J.C., Stickgold, R., Dinges, D.F. and Thompson-Schill, S.L. (2006) "Interfering with Theories of Sleep and Memory: Sleep, Declarative Memory, and Associative Interference", *Curr Biol*, 16, 1290–94.

Gaab, N., Paetzold, M., Becker, M., Walker, M.P. and Schlaug, G. (2004) "The Influence of Sleep on Auditory Learning: A Behavioral Study", *Neuroreport*, 15, 731–34.

Gais, S. and Born, J. (2004) "Declarative Memory Consolidation: Mechanisms Acting During Human Sleep", *Learn Mem*, 11, 679–85.

Huber, R., Ghilardi, M.F., Massimini, M. and Tononi, G. (2004) "Local Sleep and Learning", *Nature*, 430, 78–81.

Karni, A. and Sagi, D. (1991) "Where Practice Makes Perfect in Texture Discrimination: Evidence for Primary Visual Cortex Plasticity", *Proc Natl Acad Sci USA*, 88, 4966–70.

Karni, A., Tanne, D., Rubenstein, B.S., Askenasy, J.J. and Sagi, D. (1994) "Dependence on Rem Sleep of Overnight Improvement of a Perceptual Skill", *Science*, 265, 679–82.

Kensinger, E.A., Garoff-Eaton, R. and Schacter, D.L. (2007) "Effects of Emotion on Memory Specificity: Memory Trade-Offs Elicited by Negative Visually Arousing Stimuli", *Journal of Memory and Language*, 56, 575–91.

Mednick, S., Nakayama, K. and Stickgold, R. (2003) "Sleep-Dependent Learning: A Nap is as Good as a Night", *Nat Neurosci*, 6, 697–8.

Nishida, M. and Walker, M.P. (2007) "Daytime Naps, Motor Memory Consolidation and Regionally Specific Sleep Spindles", *PLoS ONE*, 2, e341.

Payne, J., Propper, R., Walker, M. and Stickgold, R. (2006) "Sleep Increases False Recall of Semantically Related Words in the Deese-Roediger-Mcdermott Memory Task", *Sleep*, 29, A373.

Payne, J.D., Schacter, D.L., Propper, R.E., Huang, L.W., Wamsley, E.J., Tucker, M.A., Walker, M.P. and Stickgold, R. (2009) "The Role of Sleep in False Memory Formation", *Neurobiol. Learn Mem.*

Plihal, W. and Born, J. (1997) "Effects of Early and Late Nocturnal Sleep on Declarative and Procedural Memory", *Journal of Cognitive Neuroscience*, 9, 534–47.

Roediger, H. and McDermott, K. (1995) "Creating False Memories: Remembering Words Not Presented in Lists", *Journal of Experimental Psychology: Learning, Memory, & Cognition*, 21, 803–14.

Stickgold, R., James, L. and Hobson, J.A. (2000) "Visual Discrimination Learning Requires Sleep after Training", *Nat Neurosci*, 3, 1237–8.

Stickgold, R., Whidbee, D., Schirmer, B., Patel, V. and Hobson, J.A. (2000) "Visual Discrimination Task Improvement: A Multi-Step Process Occurring During Sleep", *J Cogn Neurosci*, 12, 246–54.

Tucker, M.A., Hirota, Y., Wamsley, E.J., Lau, H., Chaklader, A. and Fishbein, W. (2006) "A Daytime Nap Containing Solely Non-Rem Sleep Enhances Declarative but Not Procedural Memory", *Neurobiology of Learning and Memory*, 86, 241–7.

Wagner, U., Gais, S. and Born, J. (2001) "Emotional Memory Formation Is Enhanced across Sleep Intervals with High Amounts of Rapid Eye Movement Sleep", *Learn Mem*, 8, 112–19.

Walker, M.P., Brakefield, T., Morgan, A., Hobson, J.A. and Stickgold, R. (2002) "Practice with Sleep Makes Perfect: Sleep-Dependent Motor Skill Learning", *Neuron*, 35, 205–11.

Walker, M.P., Brakefield, T., Seidman, J., Morgan, A., Hobson, J.A. and Stickgold, R. (2003) "Sleep and the Time Course of Motor Skill Learning", *Learn Mem*, 10, 275–84.

Walker, M.P. and Stickgold, R. (2006) "Sleep, Memory, and Plasticity", *Annu Rev Psychol*, 57, 139–66.

Yoo, S.S., Hu, P.T., Gujar, N., Jolesz, F.A. and Walker, M.P. (2007) "A Deficit in the Ability to Form New Human Memories without Sleep", *Nat Neurosci*, 10, 385–92.

Chapter 9
Factors Affecting Mnemonic Performance in a Nonhuman Primate Model of Cognitive Workload

Robert E. Hampson

Sam A. Deadwyler

Maintenance of performance following periods of extended wakefulness has become more of a rule than the exception by current operational standards. Assessment of the impact of sleep loss on cognitive function has therefore become important in several different contexts (Beaumont et al. 2001, Horne 2000, and Quigley et al. 2000). With ever increasing demands to work longer hours and more days in succession, individuals become more prone to the health disorders brought about by sleep loss (Smith et al. 2002, and Wiegmann et al. 1996). Therefore, agents that can counteract the effects of sleep deprivation are now increasing in demand, but often such agents produce their benefit accompanied by undesirable side effects (Wiegmann et al. 1996, Landolt et al. 2000, Smith et al. 1999, and Baranski and Pigeau 1997).

A critical phase of testing for such agents is development of a nonhuman model that will allow for preclinical testing of pharmaceutical agents that also mimics the conditions of sleep deprivation and cognitive load. We have developed a nonhuman primate model of cognitive workload in order to perform electrophysiological assessment of factors affecting cognition as well as to test pharmacological interventions that may improve cognitive performance under debilitating conditions (Deadwyler et al. 2007, Lindner et al. 2008). Rhesus monkeys have been trained to perform a visual delayed-match-to-sample (DMS) task using clip-art images as sample, match and distractor (nonmatch) stimuli (Hampson et al. 2004). Task parameters manipulate the number of distractor images, the delay between sample and match presentation, the amount of time that stimuli are presented, and whether the match response is to the same image as the sample (for example, "object" match) or to the same position as the sample ("spatial" match). We have shown that performance of this task is affected by the same factors which alter human cognitive performance: delay, distraction, sleep deprivation and/or fatigue, and pharmacological agents that impair cognitive decision-making (Porrino et al. 2005). We have also used the nonhuman primate model to explore different techniques for assessing performance and cognitive workload—namely behavioral, electrophysiological, occulometric, and brain metabolic assessment via positron-emission tomography. Most importantly, we have performed a comparative evaluation of pharmacological agents that may benefit cognitive performance and alleviate the cognitive impairment caused by sleep deprivation and fatigue (Porrino et al. 2005, Lindner et al. 2008). We are encouraged that results from this nonhuman primate model can be applied to human studies for development of effective countermeasures to the debilitating effects of stress and fatigue in human performance.

Nonhuman Primate Model of Cognitive Workload

Twelve rhesus macaque monkeys were tested in various facets of the study. Monkeys were trained and tested under normal conditions in a multi-image visual delayed-match-to-sample (DMS) task in which the number of images to chose from in the Match phase after a 1–30 second delay varied from two to six images (Porrino et al. 2005). The same monkeys were also subjected to a 30–36 hour sleep deprivation regimen, then tested on the full 2–6 image DMS task. Figure 9.1 illustrates the DMS task and exhibits performance curves averaged across all monkeys tested under normal (non-sleep deprived) conditions. The effect of duration of delay interval and number of images the Match phase of the DMS task is shown summed over all trials for normal (alert) sessions. The DMS performance curves at the lower right in Figure 9.1 demonstrate that overall average performance was maintained at 70–75 percent across all trials but there was a clear effect of duration of delay where performance decreased from 92 percent at short delays (1–5 seconds) to 60 percent at longer delays (25–30 seconds) ($F_{(5,1391)} = 9.24$, $p<0.001$) and that increasing the number of match choices also independently reduced overall performance ($F_{(4,1391)} = 4.93$, $p<0.001$). Effects were consistent in individual monkeys, with all showing the same effects of delay and number of images.

To further the effect of increasing difficulty and the relationship with cognitive load (Hampson et al. 2009, Deadwyler et al. 2007), trials were segregated into high and low cognitive load as illustrated in Figure 9.2. Trials with short delays and few images were classified as low cognitive load (Low Load: 1–10 seconds, 2–3 images, Figure 9.2A) while trials with long delays and many images were classified as high cognitive load (High Load: 21–30 seconds, 5–6 images, Figure 9.2A). Figure 9.2B shows DMS performance on all trials, low cognitive load trials only and high cognitive trials only in sessions consisting of a mixture of all trial types (mixed sessions) compared to performance on sessions composed exclusively of high or low cognitive load trials (exclusive sessions). Performance on high cognitive load trials was significantly reduced ($F_{(1,1391)} = 19.31$, $p<0.001$) compared to low cognitive load trials in both mixed and exclusive sessions. Moreover respective performance on high or low cognitive load trials did not differ between mixed and exclusive sessions ($F_{(2,1391)} = 0.75$ {not significant}).

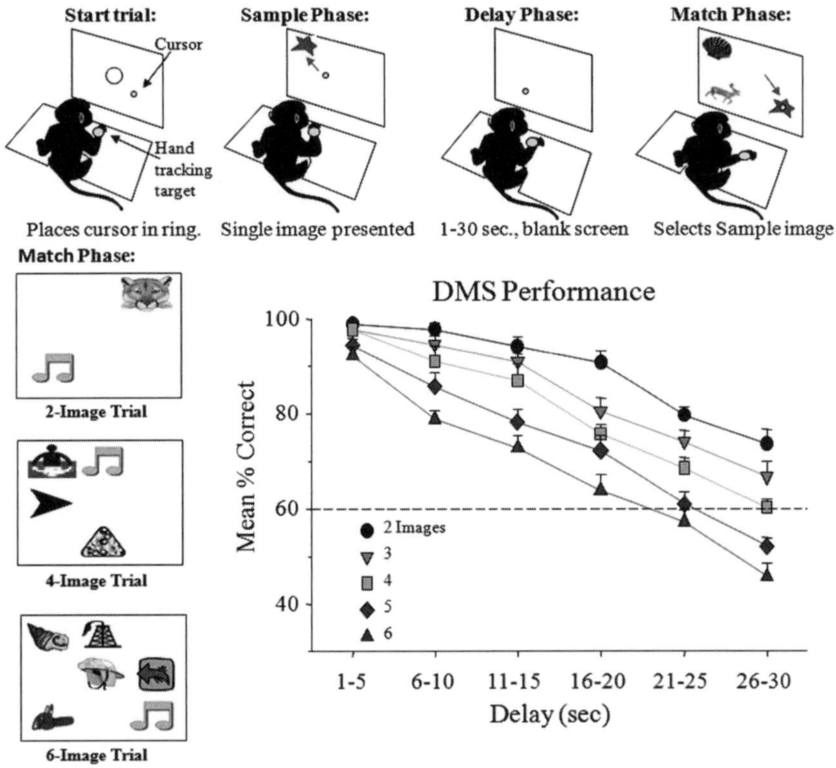

Figure 9.1 Delayed nonmatch to sample task and performance

Top: Schematic of the phases of the DMS task denoting the single image presented in the sample phase, with the response to that same image despite the presence of distractor images in the Match phase. Lower Left: Examples of 2-, 4- and 6-image Match Phase screens. Lower Right: DMS performance curves showing effect of delay and number of images averaged over 12 animals.

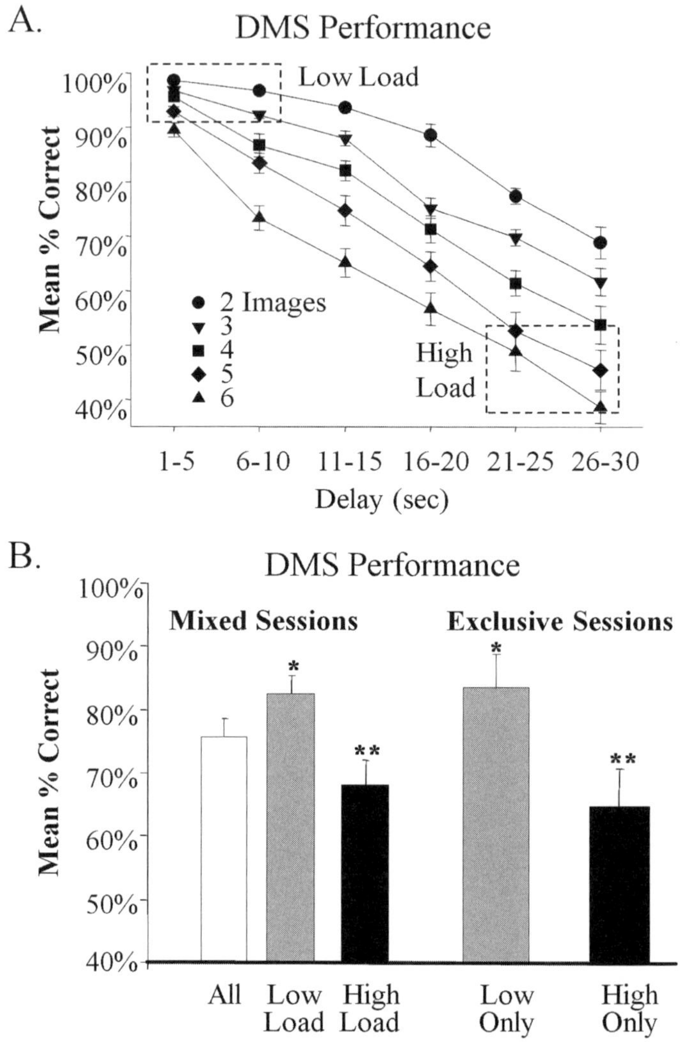

Figure 9.2 Classification of DMS trials according to cognitive workload

A: DMS performance by delay and number of Match phase images as in Figure 9.1 with high and low cognitive load indicated by dashed boxes. B: Comparison of DMS performance on high and low cognitive trials in sessions composed of mixed trial types or exclusively low/high load trials. *$p<0.01$; **$p<0.001$ compared to overall performance (All).

Figure 9.3 shows that the dramatic effect of 30–36 hours of sleep deprivation on the two monkeys' DMS performance relative to non-sleep deprived conditions. Performance was seriously altered as indicated by deficiencies at nearly all delay interval durations, number of choice images in the Match phase ($F_{(10,521)} = 4.61$, $p<0.001$). The fact that animals completed 85 percent of trials during the session and that performance was not significantly affected on the trials with no delay and only two Match phase images indicates that the sleep loss-related deficits did not result from an inability to perform the task. These performance changes were accompanied by changes in electroencephalography (EEG) and brain glucose utilization as previously reported (Porrino et al. 2005, Hampson et al. 2009) and proved to be a reliable "marker" of sleep deprived performance in both monkeys. It is also worthwhile noting that while sleep deprivation produced a slight decrease in performance on low cognitive load trials, the most profound influence was on high cognitive load trials, suggesting that the nonhuman primate DMS performance model was an effective testbed for evaluation of varying influences on cognitive workload.

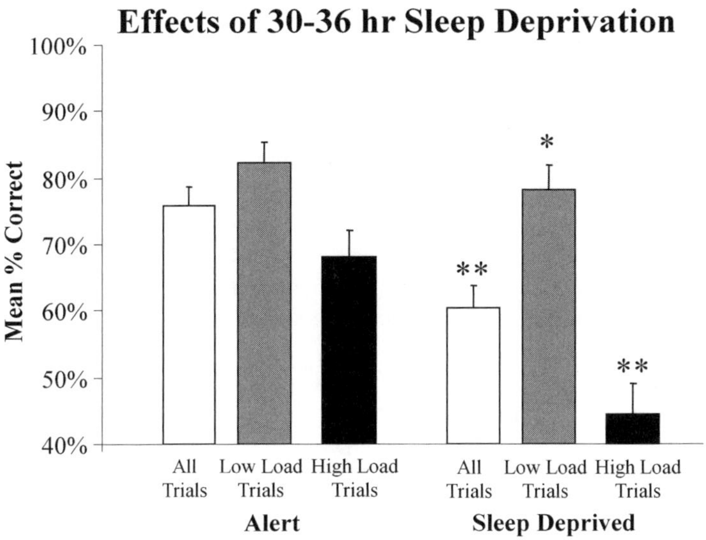

Figure 9.3 DMS performance following 30–36 hours of continuous wakefulness

Animals were deprived of rapid eye movement (REM) sleep for one night and tested in the DMS task the next day. *p<0.01, **p<0.001 compared to normal alert condition.

Pharmacological Interventions

Using the model, we were able to test various drugs shown to increase alertness to determine whether they could overcome cognitive performance deficits following sleep deprivation. Figure 9.4 compares effects of the ampakine CX717 (cortex Pharmaceuticals), amphetamine, caffeine and modafinil (Provigil) on DMS performance following sleep deprivation. When CX717 (0.8 mg/kg, intravenous) was administered to sleep deprived animals prior to performing the task, the deleterious effects on performance were markedly reversed. DMS performance following sleep deprivation was not significantly different from control (alert) sessions ($F_{(1,521)} = 3.04$ {not significant}) and even showing a slight trend toward improvement consistent CX717 effects under normal conditions, (Figure 9.4). Reversal of the debilitating effects of sleep deprivation included improved DMS performance at all delays and number of images, but in particular on longer delays (26–30 seconds, sleep deprivation, 56.2, 1.1 percent, *versus* sleep deprivation + CX717, 77.5, 0.6 per cent; $F_{4,521)} = 11.89$, $p<0.001$) and number of images (six images, sleep deprivation, 54.4, 1.5 percent *versus* sleep deprivation + CX717, 76.1, 0.9 percent; $F_{(5,521)} = 12.55$, $p<0.001$), for high cognitive load trials. The reversal of effects of sleep deprivation by CX717 was also evident in positron emission tomography (PET) scans. Sleep deprived decreased PET imaging correlates of global cerebral metabolic rate for glucose (CMR_{glc}; [^{18}F] Fluorodeoxyglucose uptake during DMS performance) in prefrontal cortex, thalamus, cingulate, and striatum, and increased CMR_{glc} in medial temporal lobe (Porrino et al. 2005). CX717 increased CMR_{glc} following sleep deprivation in prefrontal cortex, cingulate and striatum, while reducing CMR_{glc} in medial temporal lobe.

Figure 9.4 also compares the effectiveness of amphetamine, caffeine and modafinil, on the same animals under the same sleep deprivation conditions. It can be seen that amphetamine, caffeine and modafinil did not produce the same degree of reversal of DMS performance as CX717, even though all drugs enhanced high cognitive load trials, only CX717 (and amphetamine in combination with CX717) enhanced low cognitive load trials. These differential effects on performance are accompanied by differential effects on CMR_{glc}. Table 9.1 lists standard scores (Z) for region of interest (ROI) analyses of CMR_{glc} in eight brain areas previously shown to be affected by sleep deprivation (Porrino et al. 2005, and Hampson et al. 2009). These scores indicate significant changes in CMR_{glc} in those brain regions, with Z values for increase or decrease greater than 2.33 significant at the $p<0.01$ level, and Z values greater than 3.09 significant at the $p<0.001$ level. CX717 had a broader effect on the brain, even though it had no effect on thalamus. It is interesting that even though CX717 improved performance, it did not necessarily increase the observed alertness of the animals in the same manner as amphetamine and modafinil—which did alter CMR_{glc} in thalamus.

Each drug was tested for dose-dependence of effects on performance, alertness, changes in EEG and CMR_{glc}. Results reported here are based on the most effective dose that could be administered to animals without producing overt signs of overstimulation (hyperactivity, nausea, distress, disruption of recovery sleep). These results demonstrate that positive modulators of the AMPA receptor such as CX717, and possibly other members of the ampakine family (Arai et al. 2000), may be effective deterrents to the disruptive effects of sleep deprivation on cognitive performance in humans. CX717 was also effective in enhancing DMS performance under normal alert conditions, therefore its potential to reverse the detrimental effects of sleep deprivation may result from a general enhancement of performance without necessarily *reversing* alertness in the same manner as amphetamine, caffeine or modafinil. The degree to which CX717 ameliorated deficits induced by sleep deprivation was surprising and suggests that enhancement of AMPA receptor activation may be expressed more dramatically if there is underlying perturbation of brain function.

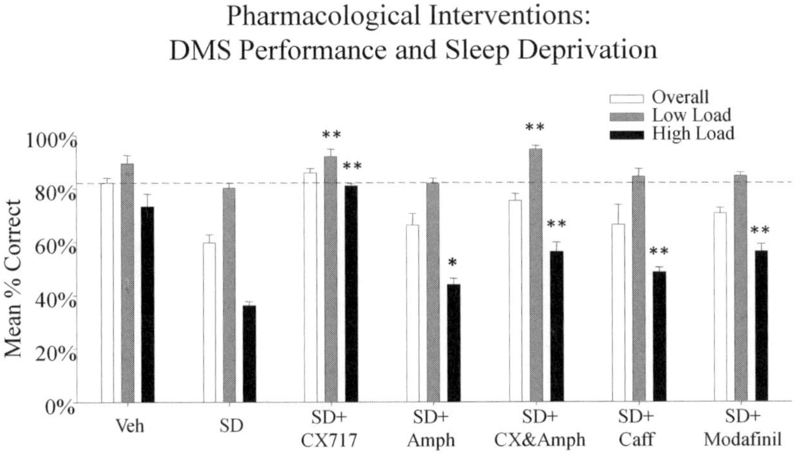

Figure 9.4 Effectiveness of various alerting agents on DMS performance impaired by sleep deprivation

Each agent was administered ten minutes before the start of the DMS session following 30–36 hours of continuous wakefulness (REM sleep deprivation). Each drug was tested in the same animals, with all animals receiving each drug for a minimum of three sessions. Horizontal dashed line shows mean percent correct for all trials (averaged over sessions and animals) for vehicle (Veh) alert controls. Sleep deprivation = Sleep deprived; CX = ampakine CX717 (Cortex Pharm., 0.8 mg/kg, iv); Amph = d-amphetamine (150 μg/kg, iv); Caff = caffeine (15 mg/kg, iv); Modafinil (10 mg/kg, iv). $^{*}p<0.01$, $^{**}p<0.001$ compared to sleep deprived.

Table 9.1 Significant PET activations during sleep-deprived DMS performance

	CX717	AMPH	CAFF	MODA
Medial temporal lobe	**4.05**	–	**2.96**	–
Dorsal striatum	3.66	4.27	4.05	**3.37**
Precuneus	**3.22**	–	–	
Prefrontal cortex	3.60	–	–	3.85
Cingulate cortex	**3.03**	–	3.38	–
Visual areas	**2.69**	**4.15**	–	5.11
Cerebellum	**2.44**	4.34	–	4.30
Thalamus	–	2.94	–	3.14
Performance deficit reversal	Full	Partial	Partial	Partial

Increased activity is underlined

Decreased activity is in bold

Z>2.33, p<0.01; Z>3.09, p<0.001

Nasally-Administered Neuroactive Peptide

The peptide agonist of the CNS orexin-A (hypocretin-1) receptor has previously been shown to increase alertness (Espana et al. 2001, Brisbare-Roch et al. 2007, and Deadwyler et al. 2007). Unfortunately, the peptide structure of Orexin makes it unsuitable for oral or intravenous administration (Deadwyler et al. 2007, Dhuria et al. 2009, and Roecker and Coleman 2008). We consequently developed a novel nasal spray application technique for delivery of orexin to animals performing the DMS task. Orexin was mixed in a saline solution and sprayed in a mist through a pulsed pressurized atomizer held directly in front of the monkey's nostrils to deliver approximately 1.0 μg orexin to each animal administered 3–5 minutes prior to testing in both the alert normal condition, as well as following 30–36 hours of sleep deprivation (Deadwyler et al. 2007). The effects of the nasal orexin-A spray on sleep-deprived sessions are shown in Figure 9.5. Nasal orexin-A had little or no effect on DMS performance under normal alert testing conditions (not shown) but markedly improved performance of the same monkeys when sleep deprived (Figure 9.5). The nasal orexin-A spray reversed the detrimental influence of sleep deprivation on low-load trials by returning performance to normal alert levels ($F_{(1,503)}=8.19$, $p<0.01$ versus sleep deprivation Saline; $F_{(1,503)}=1.14$, {not significant}versus Alert Saline). More surprisingly, however, nasal orexin-A elevated performance *above normal alert levels* on high-load trials ($F_{(1,503)}=9.07$, $p < 0.001$) which, compared to performance following control (saline) nasal spray in sleep deprived animals ($F_{(1,503)}=17.53$, $p<0.001$), constituted a considerable improvement (Figure 9.5). PET imaging revealed that orexin had similar effects on CMR_{glc} as CX717, with one key difference, an increase in thalamic activation similar to amphetamine and modafinil (Table 9.1).

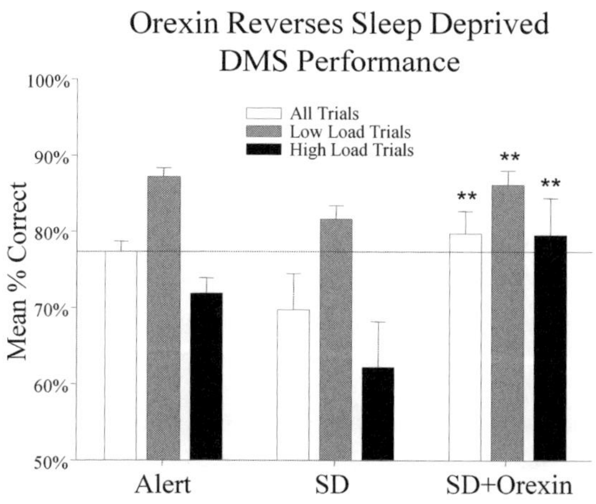

Figure 9.5 Effectiveness of orexin on DMS performance impaired by sleep deprivation

Orexin (1.0 µg) was delivered by nasal spray 3–5 minutes before start of DMS session following 30–36 hrs sleep deprivation. $^{*}p<0.01$, $^{**}p<0.001$ compared to sleep deprived.

Conclusions

Recent evidence strongly suggests that sleep deprivation may selectively downregulate critical receptor mediated cellular processes, compromising plasticity in many cortical structures (Pace-Schott and Hobson 2002, Meerlo et al. 2002, Shaffery et al. 2002, Siegel 2001, Stickgold et al. 2001, and Maquet 2001). The receptors for the excitatory neurotransmitter glutamate (both AMPA- and NMDA-type receptors) may not be affected in the same manner during sleep deprivation (Graves et al. 2001, and McDermott et al. 2003). Hence, the enhancement of AMPA receptor function in the large number of brain regions affected by sleep deprivation (Table 9.1) may be capable of reversing the deleterious effects by compensating for a possible loss of NMDA receptor function in those areas (Graves et al. 2001). Whether the effects of sleep deprivation on cognitive performance actually result from suppressed AMPA receptor mediated activity or some other receptor dependent process, compounds such as ampakines that modulate so many different glutamatergic pathways and circuits in different brain regions (Arai et al. 2000, and Danysz 2002) could provide a sound basis for agents that would counter deficits in cognition and performance provoked by sleep deprivation (Quigley et al.

2000, Drummond et al. 2000, Kingshott et al. 2000, Beaumont et al. 2001, Smith, McEvoy, and Gevins 2002, Thomas et al. 2000, Ferrara et al. 2000, and Clark, Frank, and Brown 2001). The effects of CX717 appear to be specific to preserving high levels of performance under the stressful conditions of accumulated sleep loss and provide a potential viable means of sustaining or recovering critical functions when sleep deprivation is unavoidable.

Another important receptor system involved in sleep-waking cycles is the Orexin-A/Hypocretin-1 system. The agonist peptide orexin has been shown to prevent sleep in narcoleptic dogs and has also been shown to release glutamate (Siegel 2004). Consequently, the sleep preventing effects of orexin may be potentiated by CX717. Ampakines, therefore, have the capability to positively modulate many different glutamatergic circuits and pathways in several pertinent brain regions (Lynch 2002), thereby increasing the potential to counteract cognitive deficits under normal as well as sleep deprived conditions (Drummond and Brown 2001, Smith, McEvoy, and Gevins 2002, Thomas et al. 2000, and Ferrara et al. 2000).

Other compounds that have been utilized to combat the effects of sleep deprivation, including psychostimulants (amphetamine), caffeine and modafinil (Beaumont et al. 2001, Wiegmann et al. 1996, Wesensten et al. 2002, Martinez-Gonzalez et al. 2004, and Wesensten et al. 2004) act through different, non-AMPA receptor mediated, cell signaling pathways. The usefulness of these agents, however, may be limited due to their potential for addiction and/or potent stimulant actions which can distort cognitive and sensory processes at doses required to counteract the effects of sleep deprivation (Beaumont et al. 2001, Wiegmann et al. 1996, Wesensten et al. 2002, Martinez-Gonzalez et al. 2004, and Wesensten et al. 2004). In addition, even though such compounds may alter transmission at critical synapses engaged in different types of cognitive functions nonetheless, they must do so through cooperative activation of glutamatergic synapses (Johnson et al. 1999, and Goff et al. 2001).

The current findings validate the nonhuman primate model of behavioral performance and cognitive workload by providing further evidence for the involvement of the both the NMDA and orexin receptor systems in the regulation of important cognitive processes that are affected by physiological perturbations such as sleep deprivation. The results have demonstrated that the task can identify conditions of high and low cognitive workload as well as demonstrating the effectiveness of glutamatergic and orexinergic modulators to reactivate specific brain mechanisms required for optimal performance in a task requiring both attention and working memory (Hampson et al. 2004). Although it is possible that the reversal of the effects on sleep deprivation could be mediated via peripheral actions of nasal orexin-A on olfactory nerves or other neural processes (van den Pol 1998, Kilduff and Peyron 2000), the likelihood that such actions would affect the same brain regions (Figure 9.6) as the ampakine CX717 (Porrino et al. 2005) tested in the same behavioral paradigm, is remote. Such investigation therefore also reveals new opportunities to utilize pharmaceutical alerting agents and other neuroactive peptides in applications related to sleep, as well as in other types of brain disorders, via a much less invasive nasal route of administration.

References

Arai, A.C., Kessler, M., Rogers, G. and Lynch, G. (2000) "Effects of the Potent Ampakine Cx614 on Hippocampal and Recombinant Ampa Receptors: Interactions with Cyclothiazide and Gyki 52466", *Mol Pharmacol*, 58, 802–13.

Baranski, J.V. and Pigeau, R.A. (1997) "Self-Monitoring Cognitive Performance During Sleep Deprivation: Effects of Modafinil, D-Amphetamine and Placebo", *J Sleep Res*, 6, 84–91.

Beaumont, M., Batejat, D., Pierard, C., Coste, O., Doireau, P., Van Beers, P., Chauffard, F., Chassard, D., Enslen, M., Denis, J.B. and Lagarde, D. (2001) "Slow Release Caffeine and Prolonged (64-H) Continuous Wakefulness: Effects on Vigilance and Cognitive Performance", *J Sleep Res*, 10, 265–76.

Brisbare-Roch, C., Dingemanse, J., Koberstein, R., Hoever, P., Aissaoui, H., Flores, S., Mueller, C., Nayler, O., van Gerven, J., de Haas, S.L., Hess, P., Qiu, C., Buchmann, S., Scherz, M., Weller, T., Fischli, W., Clozel, M. and Jenck, F. (2007) "Promotion of Sleep by Targeting the Orexin System in Rats, Dogs and Humans", *Nat Med*, 13, 150–55.

Clark, C.P., Frank, L.R. and Brown, G.G. (2001) "Sleep Deprivation, Eeg, and Functional Mri in Depression: Preliminary Results", *Neuropsychopharmacology*, 25, S79–84.

Danysz, W. (2002) "Cx-516 Cortex Pharmaceuticals", *Curr Opin Investig Drugs*, 3, 1081–8.

Deadwyler, S.A., Porrino, L., Siegel, J.M. and Hampson, R.E. (2007) "Systemic and Nasal Delivery of Orexin-a (Hypocretin-1) Reduces the Effects of Sleep Deprivation on Cognitive Performance in Nonhuman Primates", *J Neurosci*, 27, 14239–47.

Dhuria, S.V., Hamson, L.R. and Frey, W.H., 2nd (2009) "Intranasal Drug Targeting of Hypocretin-1 (Orexin-a) to the Central Nervous System", *J Pharm Sci*, 98, 2501–15.

Drummond, S.P. and Brown, G.G. (2001) "The Effects of Total Sleep Deprivation on Cerebral Responses to Cognitive Performance", Neuropsychopharmacology, 25, S68–73.

Drummond, S.P., Brown, G.G., Gillin, J.C., Stricker, J.L., Wong, E.C. and Buxton, R.B. (2000) "Altered Brain Response to Verbal Learning Following Sleep Deprivation", *Nature*, 403, 655–7.

Espana, R.A., Baldo, B.A., Kelley, A.E. and Berridge, C.W. (2001) "Wake-Promoting and Sleep-Suppressing Actions of Hypocretin (Orexin): Basal Forebrain Sites of Action", *Neuroscience*, 106, 699–715.

Ferrara, M., De Gennaro, L., Casagrande, M. and Bertini, M. (2000) "Selective Slow-Wave Sleep Deprivation and Time-of-Night Effects on Cognitive Performance Upon Awakening", *Psychophysiology*, 37, 440–46.

Goff, D.C., Leahy, L., Berman, I., Posever, T., Herz, L., Leon, A.C., Johnson, S.A. and Lynch, G. (2001) "A Placebo-Controlled Pilot Study of the Ampakine Cx516 Added to Clozapine in Schizophrenia", *J Clin Psychopharmacol*, 21, 484–7.

Graves, L., Pack, A. and Abel, T. (2001) "Sleep and Memory: A Molecular Perspective", *Trends Neurosci*, 24, 237–43.

Hampson, R.E., Espana, R.A., Rogers, G.A., Porrino, L.J. and Deadwyler, S.A. (2009) "Mechanisms Underlying Cognitive Enhancement and Reversal of Cognitive Deficits in Nonhuman Primates by the Ampakine Cx717", *Psychopharmacology (Berl)*, 202, 355–69.

Hampson, R.E., Pons, T.P., Stanford, T.R. and Deadwyler, S.A. (2004) "Categorization in the Monkey Hippocampus: A Possible Mechanism for Encoding Information into Memory", *Proc Natl Acad Sci USA*, 101, 3184–9.

Horne, J. (2000) "Neuroscience. Images of Lost Sleep", *Nature*, 403, 605–6.

Johnson, S.A., Luu, N.T., Herbst, T.A., Knapp, R., Lutz, D., Arai, A., Rogers, G.A. and Lynch, G. (1999) "Synergistic Interactions between Ampakines and Antipsychotic Drugs", *J Pharmacol Exp Ther*, 289, 392–7.

Kilduff, T.S. and Peyron, C. (2000) "The Hypocretin/Orexin Ligand-Receptor System: Implications for Sleep and Sleep Disorders", *Trends Neurosci*, 23, 359–65.

Kingshott, R.N., Cosway, R.J., Deary, I.J. and Douglas, N.J. (2000) "The Effect of Sleep Fragmentation on Cognitive Processing Using Computerized Topographic Brain Mapping", *J Sleep Res*, 9, 353–7.

Landolt, H.P., Finelli, L.A., Roth, C., Buck, A., Achermann, P. and Borbely, A.A. (2000) "Zolpidem and Sleep Deprivation: Different Effect on Eeg Power Spectra", *J Sleep Res*, 9, 175–83.

Lindner, M.D., McArthur, R.A., Deadwyler, S.A., Hampson, R.E., and Tariot, P.N. "Development, Optimization and Use of Preclinical Behavioral Models to Maximize the Productivity of Drug Discovery for Alzheimer's Disease" in *Animal and Translational Models of Behavioral Disorders, Vol. 2—Neurologic Disorders*, McArthur, R.A., Borsini, F., (eds), Academic Press, New York, 2008.

Lynch, G. (2002) "Memory Enhancement: The Search for Mechanism-Based Drugs", *Nat Neurosci*, 5 Suppl, 1035–8.

Maquet, P. (2001) "The Role of Sleep in Learning and Memory", *Science*, 294, 1048–52.

Martinez-Gonzalez, D., Obermeyer, W., Fahy, J.L., Riboh, M., Kalin, N.H. and Benca, R.M. (2004) "Rem Sleep Deprivation Induces Changes in Coping Responses That Are Not Reversed by Amphetamine", *Sleep*, 27, 609–17.

McDermott, C.M., LaHoste, G.J., Chen, C., Musto, A., Bazan, N.G. and Magee, J.C. (2003) "Sleep Deprivation Causes Behavioral, Synaptic, and Membrane Excitability Alterations in Hippocampal Neurons", *J Neurosci*, 23, 9687–95.

Meerlo, P., Koehl, M., van der Borght, K. and Turek, F.W. (2002) "Sleep Restriction Alters the Hypothalamic-Pituitary-Adrenal Response to Stress", *J Neuroendocrinol*, 14, 397–402.

Pace-Schott, E.F. and Hobson, J.A. (2002) "The Neurobiology of Sleep: Genetics, Cellular Physiology and Subcortical Networks", *Nat Rev Neurosci*, 3, 591–605.

Porrino, L.J., Daunais, J.B., Rogers, G.A., Hampson, R.E. and Deadwyler, S.A. (2005) "Facilitation of Task Performance and Removal of the Effects of Sleep Deprivation by an Ampakine (Cx717) in Nonhuman Primates", *PLoS Biol*, 3, e299.

Quigley, N., Green, J.F., Morgan, D., Idzikowski, C. and King, D.J. (2000) "The Effect of Sleep Deprivation on Memory and Psychomotor Function in Healthy Volunteers", *Hum Psychopharmacol*, 15, 171–7.

Roecker, A.J. and Coleman, P.J. (2008) "Orexin Receptor Antagonists: Medicinal Chemistry and Therapeutic Potential", *Curr Top Med Chem*, 8, 977–87.

Shaffery, J.P., Sinton, C.M., Bissette, G., Roffwarg, H.P. and Marks, G.A. (2002) "Rapid Eye Movement Sleep Deprivation Modifies Expression of Long-Term Potentiation in Visual Cortex of Immature Rats", *Neuroscience*, 110, 431–43.

Siegel, J.M. (2001) "The Rem Sleep-Memory Consolidation Hypothesis", *Science*, 294, 1058–63.

Siegel, J.M. (2004) "The Neurotransmitters of Sleep", *J Clin Psychiatry*, 65 Suppl 16, 4–7.

Smith, G.S., Reynolds, C.F., 3rd, Pollock, B., Derbyshire, S., Nofzinger, E., Dew, M.A., Houck, P.R., Milko, D., Meltzer, C.C. and Kupfer, D.J. (1999) "Cerebral Glucose Metabolic Response to Combined Total Sleep Deprivation and Antidepressant Treatment in Geriatric Depression", *Am J Psychiatry*, 156, 683–9.

Smith, M.E., McEvoy, L.K. and Gevins, A. (2002) "The Impact of Moderate Sleep Loss on Neurophysiologic Signals During Working-Memory Task Performance", *Sleep*, 25, 784–94.

Stickgold, R., Hobson, J.A., Fosse, R. and Fosse, M. (2001) "Sleep, Learning, and Dreams: Off-Line Memory Reprocessing", *Science*, 294, 1052–7.

Thomas, M., Sing, H., Belenky, G., Holcomb, H., Mayberg, H., Dannals, R., Wagner, H., Thorne, D., Popp, K., Rowland, L., Welsh, A., Balwinski, S. and Redmond, D. (2000) "Neural Basis of Alertness and Cognitive Performance Impairments During Sleepiness. I. Effects of 24 H of Sleep Deprivation on Waking Human Regional Brain Activity", *J Sleep Res*, 9, 335–52.

van den Pol, A.N., Gao, X.B., Obrietan, K., Kilduff, T.S. and Belousov, A.B. (1998) "Presynaptic and Postsynaptic Actions and Modulation of Neuroendocrine Neurons by a New Hypothalamic Peptide, Hypocretin/Orexin", *J Neurosci*, 18, 7962–71.

Wesensten, N.J., Belenky, G., Kautz, M.A., Thorne, D.R., Reichardt, R.M. and Balkin, T.J. (2002) "Maintaining Alertness and Performance During Sleep Deprivation: Modafinil Versus Caffeine", *Psychopharmacology (Berl)*, 159, 238–47.

Wesensten, N.J., Belenky, G., Thorne, D.R., Kautz, M.A. and Balkin, T.J. (2004) "Modafinil vs. Caffeine: Effects on Fatigue During Sleep Deprivation", *Aviat Space Environ Med*, 75, 520–25.

Wiegmann, D.A., Stanny, R.R., McKay, D.L., Neri, D.F. and McCardie, A.H. (1996) "Methamphetamine Effects on Cognitive Processing During Extended Wakefulness", *Int J Aviat Psychol*, 6, 379–97.

PART 3
Cognition During Stress and Anxiety

Chapter 10

Systems Neuroscience Approaches to Measure Brain Mechanisms Underlying Resilience—Towards Optimizing Performance

Martin P. Paulus

Alan N. Simmons

Eric G. Potterat

Karl F. Van Orden

Judith L. Swain

Introduction

From a systems neuroscience perspective, optimal performance under extreme conditions can be conceptualized as goal-oriented task completion during high demand contexts. This conceptualization highlights the importance of cognitive control, affect management, and learning or adaptation of these systems in extreme environments. An experimental understanding of this conceptualization can be obtained by postulating two adaptive top-down cognitive processes: (1) feedback/integration of an adverse outcome (cognitive appraisal); and (2) anticipatory adaptation. These two processes combine to maximize the information extracted from environmental cues to optimize future responses. The top-down modulatory ability is fundamentally related to the cognitive appraisal notion introduced above, and to learning associations between stimuli and future pleasant or aversive outcomes. For example, the rate of reward learning depends on the discrepancy between the actual occurrence of reward and the predicted occurrence of reward, the so-called "reward prediction error" (Schultz, Dayan, and Montague 1997). Below we briefly review aspects of stress, the key neural substrates in performing under stressful conditions, and the proposed role of two brain areas that may contribute to optimal performance in extreme conditions.

We have developed a preliminary model of optimal performance in extreme environments (Paulus et al. in press) that starts with the observation that these environments exert profound interoceptive effects. Interoception is (a) sensing the physiological condition of the body (Craig 2002), (b) representing the internal state

(Craig 2009) within the context of ongoing activities, and (c) initiating motivated action to homeostatically regulate the internal state (Craig 2007). Interoception includes a range of sensations such as pain (LaMotte et al. 1982), temperature (Craig and Bushnell 1994), itch (Schmelz et al. 1997), tickle (Lahuerta et al. 1990), sensual touch (Vallbo et al. 1995, Olausson et al. 2002), muscle tension (Light and Perl 2003), air hunger (Banzett et al. 2000), stomach pH (Feinle 1998), and intestinal tension (Robinson et al. 2005), which together provide an integrated sense of the body's physiological condition (Craig 2002). These sensations travel via small-diameter primary afferent fibers, which eventually reach the anterior insular cortex for integration (Craig 2003b).

The interoceptive system provides this information to (1) systems that monitor value and salience (orbitofrontal cortex and amygdala); (2) are important for evaluating reward (ventral striatum/extended amygdala); and (3) are critical for cognitive control processes (anterior cingulate). Moreover, the more anterior the representation of the interoceptive state within the insular cortex the more "textured", multimodal, and complex the information that is being processed, due to the diverse cortical afferents to the mid and anterior insula. We have hypothesized that the anterior insula not only receives interoceptive information but is also able to generate a predictive model that provides the individual with a signal of how the body will feel (Paulus and Stein 2006) similar to the "as if" loop in the Damasio somatic marker model (Damasio 1994). The interoceptive information is thus "contextualized", meaning that it is brought in relation to other ongoing affective, cognitive, or experiential processes, in relation to the homeostatic state of the individual, and is used to initiate new or modify ongoing actions aimed at maintaining the individual's homeostatic state.

In this fashion interoceptive stimuli can generate an urge to act. Thus we propose the following process hypotheses: (1) individuals who are optimal performers have developed a well "contextualized" internal body state that can quickly and effectively adapt to conditional changes and extremes. In contrast, sub-optimal performers either receive interoceptive information that is too strong or too weak to adequately plan or execute appropriate actions. As a consequence, there is a mismatch between the experienced body state and the necessary action to maintain homeostasis. Therefore, a neural systems model of optimal performance in extreme environments needs to include brain structures that are able to process cognitive conflict and perturbation of the homeostatic balance, for example the anterior cingulate and insular cortex. Ultimately, engagement of these brain structures is likely to be predictive of performance and may also be used as an indicator of efficacy of an intervention.

A high degree of resiliency is required to perform optimally in changing and challenging environments. Resilience refers to (1) the ability to cope effectively with stress and adversity and (2) the positive growth following homeostatic disruption (Richardson 2002) and is an important psychological construct to examine how individuals respond to challenging situations and stay mentally and physically healthy in the process (Tugade, Fredrickson, and Barrett 2004) The ability to regulate and generate positive emotions plays an important role in the development of coping strategies when confronted with a negative event (Bonanno 2004). In particular, resilient individuals often generate positive emotions in order to rebound from stressful encounters (Tugade et al. 2004). Nevertheless, the experimental assessment of resilience is challenging and requires novel behavioral and neural systems techniques (Charney 2006).

Resilience is a complex and possibly multi-dimensional construct (Luthar, Cicchetti, and Becker 2000). It includes trait variables such as temperament and personality as well as cognitive functions such as problem solving that may work together for an individual to adequately cope with traumatic events (Campbell-Sills, Cohan, and Stein 2006). Here, we focus on resilience in terms of a process through which individuals successfully cope with (and bounce back from) stress. For instance, after being fired from a job, a resilient individual adopts a proactive style in improving his job hunting and work performance. In contrast a less resilient individual may adopt a simple recovery from insult style where job loss causes a period of initial depressive mood followed by a return to affective baseline without an attempt to modify habitual coping mechanisms.

The current study was aimed to show that resilience, which is a critical characteristic of optimal performance in extreme environments, has significant effects on brain structures that are thought to be important for such performance. We used a task of emotion face assessment that we have previously shown to be sensitive to levels of trait anxiety (Stein et al. 2007), can be modulated by anti-anxiety drugs (Paulus et al. 2005), and is well known to be sensitive to genetic differences across individuals (Hariri et al. 2002). As elaborated earlier, we hypothesize that limbic and paralimbic structures play an important role in helping individuals adjust to extreme conditions. Thus, we hypothesize that activation of the amygdala and insular cortex is critically modulated by the level of resilience. In particular, if the anterior insular plays an important role in helping to predict perturbations in the internal body state, one would hypothesize that greater activation in this structure is associated with better resilience. Moreover, if one assumes that the amygdala is important in assessing salience in general and the potential of an aversive impact in particular, one would hypothesize that greater resilience is associated with relatively less activation in the amygdala during emotion face processing. The results indeed demonstrate that activation in limbic and paralimbic areas of the brain during face emotion processing is modulated by the level of resilience.

Methods

Participants

This study was approved by the University of California San Diego (UCSD) and the San Diego State University (SDSU) Institutional Review Boards and all subjects signed informed consent. Initially, a sample of SDSU undergraduate psychology students participated in mass screening using the Spielberger Trait Anxiety Questionnaire (Spielberger 1983). All subjects were subsequently interviewed with a structured diagnostic interview (SCID) (Fist et al. 1995), modified to enable us to document the presence of sub-threshold anxiety and mood disorders. Only subjects who did not have a DSM-IV (American Psychiatric Association 1994) diagnosis were included in this study. Twenty-six subjects were studied, 17 females and nine males. The average age of the subjects was 19.0 (18–26) years old and they had an average of 12.8 (11–15) years of education. All subjects were trained to perform the emotion face-processing task prior to testing during fMRI scanning and received $50 for participation. No restrictions were placed on the consumption of caffeine-containing beverages; none of the subjects were smokers.

Measures

Resiliency

Connor-Davidson Resilience Scale (CD-RISC; (Connor & Davidson 2003)). The CD-RISC is a 25-item scale that measures the ability to cope with stress and adversity. Items include: "I am able to adapt when changes occur," "I tend to bounce back after illness, injury, or other hardships," and "I am able to handle unpleasant or painful feelings like sadness, fear, and anger." Respondents rate items on a scale from zero ("not true at all") to four ("true nearly all the time"). Prior analyses of the original CD-RISC in general population, primary care, psychiatric outpatient, and clinical trial samples support its internal consistency, test-retest reliability, and convergent and divergent validity (Connor et al. 2003). The CD-RISC also was shown to moderate the relationship between retrospective reports of childhood maltreatment and current psychiatric symptoms (Campbell-Sills, Cohan, and Stein 2006) and CD-RISC scores have been suggested to increase following treatments hypothesized to enhance resilience (Davidson et al. 2005).

Task

During fMRI, each subject was tested on a slightly modified (Paulus et al. 2005) version of the emotion face assessment task (Hariri et al. 2002, 2005). During each five-second trial, a subject is presented with a target face (on the top of the computer screen) and two probe faces (on the bottom of the screen) and is instructed to match the probe with the same emotional expression to the target by pressing the left or right key on a button box. A block consists of six consecutive trials where the target face is angry, fearful, or happy. During the sensorimotor control task subjects were presented with five-second trials of either wide or tall ovals or circles in an analogous configuration and instructed to match the shape of the probe to the target. Each block of faces and of the sensorimotor control task was presented three times in a pseudo-randomized order. A fixation cross lasting eight seconds was interspersed between each block presented at the beginning and end of the task (resulting in 14 fixation periods). For each trial, response accuracy and reaction time data were obtained. There were 18 trials (three blocks of six trials) for each face set as well as for shapes. The whole task lasted 512 seconds (matching the scan length).

Analysis

Acquisition of images: All scans were performed on a 3T General Electric CXK4 Magnet at the UCSD Keck Imaging Center, which is equipped with eight high bandwidth receivers that allow for shorter read-out times and reduced signal distortions and ventromedial signal dropout. Each one hour session consisted of a three-plane scout scan (ten seconds), a standard anatomical protocol, for example, a sagittally acquired spoiled gradient recalled (SPGR) sequence (FOV 25 cm; matrix: 192x256; 172 sagittally acquired slices thickness: 1 mm; TR: 8ms; TE: 3 msec; flip angle =12°). We used an 8-channel brain array coil to axially acquire T2*-weighted echo-planar images (EPI) with the following parameters: FOV 230 mm, 64 x 64 matrix; 30 2.6 mm thick slices; 1.4 mm gap; TR = 2000ms, TE = 32 ms, flip angle = 90°.

Image Analysis Pathway

The basic structural and functional image processing were conducted with the Analysis of Functional Neuroimages (AFNI) software package (Cox 1996). A multivariate regressor approach detailed below was used to relate changes in EPI intensity to differences in task characteristics (Haxby, Hoffman, and Gobbini 2000). EPI images were co-registered using a 3D-coregistration algorithm (Eddy, Fitzgerald, and Noll 1996) that has been developed to minimize the amount of image translation and rotation relative to all other images. Six motion parameters were obtained across the time series for each subject. Three of these motion parameters were used as regressors to adjust EPI intensity changes due to motion artifacts. All slices of the EPI scans were temporally aligned following registration to ensure different relationships with the regressors are not due to the acquisition of different slices at different times during the repetition interval.

Multiple Regressor Analyses

Four orthogonal regressors of interest were: (1) happy, (2) angry, (3) fearful, (4) circle/oval (that is, shape) sensorimotor condition. These 0-1 regressors were convolved with a gamma variate function (Boynton et al. 1996) modeling a prototypical hemodynamic response (6–8 second delay (Fristonet al. 1995)) and to account for the temporal dynamics of the hemodynamic response (typically

12–16 seconds) (Cohen, 1997). The convolved time series was normalized and used as a regressor of interest. A series of regressors of interest and the motion regressors were entered into the AFNI program, 3-D Deconvolve, to determine the height of each regressor for each subject. The key measure is the voxel-wise normalized relative signal change (or percent signal change for short), which is obtained by dividing the regressor coefficient by the sum of the zero-order regressor and the mean first-order regressor. Spatially smoothed (4 mm full width half maximum Gaussian filter) percent signal change data were transformed into Talairach coordinates based on the anatomical MR image, which is transformed manually in AFNI.

Second order analyses: The voxel-wise Talairach-transformed percent signal change data was the main dependent measure and was subjected to a multivariate regression analysis to determine the effect of resiliency of brain activation during the emotion processing task. A threshold adjustment method based on Monte-Carlo simulations was used to guard against identifying false positive areas of activation (Forman et al. 1995). Based on previous studies and simulations implemented in the AFNI program AlphaSim, it was determined that a voxel-wise a priori probability of 0.05 will result in a whole-brain corrected cluster-wise activation probability of 0.05, if a minimum volume of 1024μl and a connectivity radius of 4.0 mm is considered.

Anatomically-Constrained Functional Regions of Interest

We have developed and extended an anatomically-constrained functional region of interest (ROI) approach to test the proposed hypothesis about amygdala and insular cortex functioning. This approach was derived from our work showing amygdala activation attenuation with a benzodiazepine anxiolytic drug (Paulus et al. 2005). Moreover, this approach is supported by findings by other groups showing greatest test-retest reliability using this approach, compared to using a pure anatomical ROI or a voxel-by-voxel approach (Johnstone et al. 2005). We have extended this approach to use a probability mask of the insular cortex. The estimated individual insula data were normalized to Talairach coordinates and reprocessed in AFNI. Finally, a group insula probability mask was created in Talairach space based on the Talairach Atlas using the AFNI program 3dcalc.

Statistics

All behavioral analyses were carried out with SPSS 10.0 (Norusis 1990). A repeated measures multivariate ANOVA, with face type (angry, fearful, happy) as the within-subjects factor, was used to analyze the behavioral measures and neural activation patterns. To relate resiliency to the activation patterns during the emotion face processing task, we conducted voxelwise multiple linear regression analyses with gender and resiliency as independent measures and the percent signal difference between faces and the sensorimotor control condition as the dependent measure using the l m module of the statistical programming language R (http://cran.r-project.org/).

Results

Task-Related Activation

Activation during the emotion face assessment task included both limbic and paralimbic structures including bilateral insula, amygdala, and also the fusiform gyrus (data not shown).

Resiliency-Related Activation

Three main areas showed resilience-related activation during the emotion face assessment task. First, the ventromedial prefrontal cortex showed a significant inverse relationship between the amount of activation during emotion face processing and levels of resilience ($F(2,23) = 5.39$, $p = 0.012$, $r^2 = 0.319$).

There was no significant effect of gender ($p = 0.58$) on this degree of activation in this area. Moreover, when examining the different face types (anger, fear or happy), there was no significant resiliency by face-type interaction (Figure 10.1, $F(2,22) = 0.526$, $p = 0.59$). Thus, processing a greater level of resilience was associated similarly with less activation in this area when presented with all face types. Second, the right anterior insular cortex showed a significant positive relationship between the amount of activation during emotion face processing and levels of resilience (Figure 10.2, $F(2,23) = 6.16$, $p = 0.007$, $r^2 = 0.349$).

Similarly, there was no significant effect of gender ($p = 0.71$). Moreover, when examining the different face types (anger, fear, or happy), there was a trend in resiliency by face-type interaction ($F(2,22) = 3.365$, $p = 0.053$). Specifically, processing a greater level of resilience was associated with greater activation in this area when presented with angry faces in particular. Third, the right amygdala showed a significant positive relationship between the amount of activation during emotion face processing and levels of resilience (Figure 10.3, $F(2,23) = 4.79$, $p = 0.018$, $r^2 = 0.294$). Similarly, there was no significant effect of gender ($p = 0.85$). Moreover, when examining the different face types (anger, fear, or happy), there was no resiliency by face-type interaction ($F(2,22) = 1.852$, $p = 0.181$). Thus, similar to the findings in the ventromedial prefrontal cortex, greater resiliency was associated with attenuated activation in the amygdala irrespective of the face type.

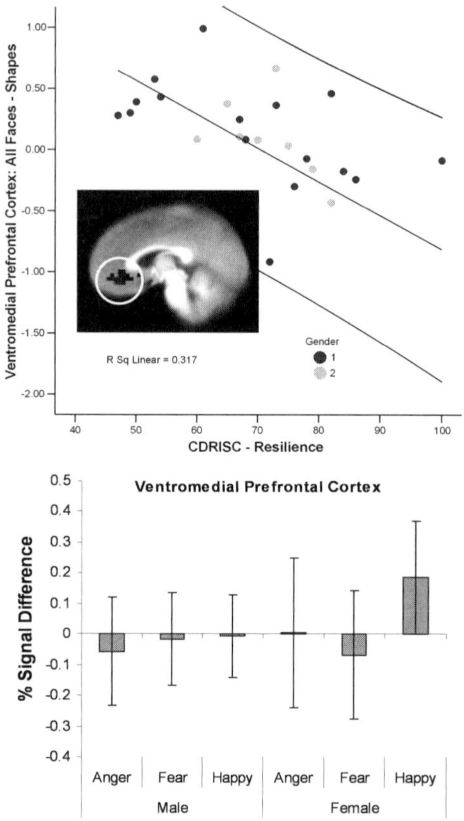

Figure 10.1 Resiliency-related brain activation in the ventromedial prefrontal cortex

Top—scatter plot of levels of resilience versus percent signal difference in the ventromedial prefrontal cortex during the face processing relative to the sensorimotor control condition. More resilience was associated with relatively less activation in this area. Bottom—average activation for different target faces for male and female participants, respectively.

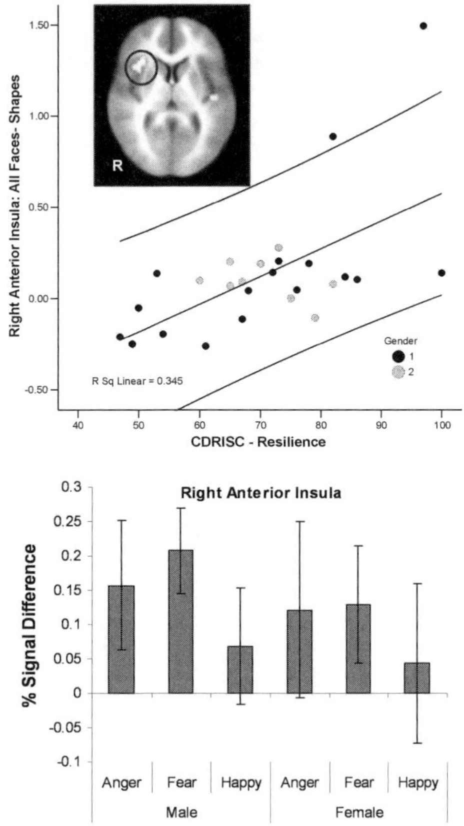

Figure 10.2 Resiliency-related brain activation in the right anterior insular cortex

Top—scatter plot of resilience versus percent signal difference in right anterior insular cortex. More resilience was associated with relatively greater activation in the insular cortex. Bottom—average activation for different target faces for male and female participants, respectively.

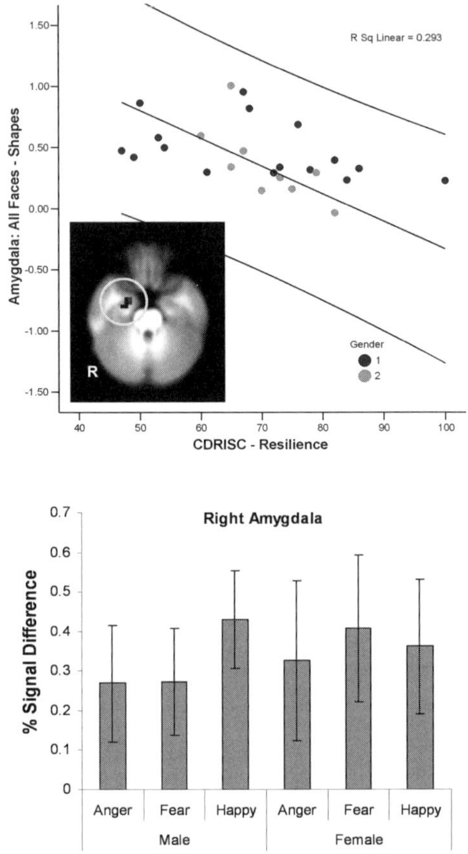

Figure 10.3 Resiliency-related brain activation in the amygdala

Top—scatter plot of resilience versus percent signal difference in right amygdala. More resilience was associated with relatively less activation in the amygdala. Bottom—average activation for different target faces for male and female participants, respectively.

Discussion

The main result of this investigation was that activation in limbic and paralimbic areas of the brain during face emotion processing are modulated by the level of resilience. In particular, greater resilience is associated with less activation in the ventromedial prefrontal cortex and amygdala, but more activation in the right anterior insular cortex. Given the notion that resilience is important for optimal performance in extreme environments, and given the finding that resilience is related to activation in brain areas, we had hypothesized the level of activation of these areas to be important for optimal performance. We offer this result as an important step to linking brain activation to optimal performance. Moreover, the involvement of the insular cortex supports our general notion that this brain structure may be critically involved in assessing ongoing internal body states as they relate to challenges in the outside world.

The insula, reviewed by (Augustine 1996, 1985), is a paralimbic structure, which constitutes the invaginated portion of the cerebral cortex, forming the base of the sylvian fissure, and is considered limbic sensory cortex by some investigators (Craig 2003a). A central insular sulcus divides the insula into two portions, the anterior and posterior insula. The anterior insula is strongly connected to different parts of the frontal lobe, whereas the posterior insula is connected to both the parietal and temporal lobes (Ture et al. 1999). The columnar organization of the insular cortex shows a highly organized anterior inferior to posterior superior gradient (for example, see Mesulam and Mufson 1982). Specifically, whereas posterior insular is characterized by a granular cortical architecture, the anterior inferior insula has an agranular columnar organization, that is, it lacks layer 4 granular cells. This type of transition is found in other parts of the brain whenever cortical representations are based on modulatory or selective feedback circuits (Shipp 2005).

Spindle cells within the anterior insular—orbitofrontal transition region (Nimchinsky et al. 1999) may be the cellular substrate underlying the possibility of widespread cortical integration. Activation of insular cortex has been reported in a number of processes including pain (Tracey et al. 2000), interoceptive (Critchley et al. 2004), emotion-related (Phan et al. 2002), cognitive (Huettel et. al 2004), and social processes (Eisenberger et al. 2003). Moreover, we have shown that the insular cortex is an important structure for processing the anticipation of aversive emotional states (Simmons et al. 2004; Simmons et al. 2006; Simmons et al. 2008), risk-taking (Paulus et al. 2003), and decision-making (Paulus 2007). In reward-related processes the insular cortex is important for subjective feeling states and interoceptive awareness (Craig 2002, Critchley et al. 2004) and has been identified as taking part in inhibitory processing with the middle and inferior frontal gyri, frontal limbic areas, and the inferior parietal lobe (Garavan, Ross, and Stein 1999).

Resilient individuals are able to generate positive emotions (Arce et al. 2008) to help them cope with extreme situations (Tugade et al. 2004). According to Fredrickson's broaden-and-build theory, positive emotions facilitate enduring personal resources and broaden one's momentary thought of action repertoire (Fredrickson 2004). That is, positive emotions broaden one's awareness and encourage novel, varied, and exploratory thoughts and actions, which in turn build skills and resources. For example, experiencing a pleasant interaction with a person you asked for directions turns, over time, into a supportive friendship. Furthermore, positive emotions help resilient individuals achieve effective coping (Werner and Smith 1992), serving to moderate stress reactivity and mediate stress recovery (Ong et al. 2006).

We suggest individuals who score high on self-reported resilience may be more likely to engage the insular cortex when processing salient information and are able to generate a body prediction error that enables them to adjust more quickly to different external demand characteristics. In turn, a more adaptive adjustment is thought to result in a more positive view of the world, and this capacity helps maintain homeostasis. This positive bias during emotion perception may provide the rose-colored glasses that resilient individuals use to interpret the world and to achieve effective ways to bounce back from adversity (Bonanno 2004) and maintain wellness.

In a recent study (Simmons et al. 2009), it was found that individuals with a prior history of significant emotional trauma, were less able to interoceptively adapt to changing interoceptive states, as evidenced by reduced anterior insula activation, although they where cognitively engaged with the state change, as evidenced by dorsal lateral prefrontal activation. These findings provide further evidence that those with reduced resilience, either prior to or following extreme stress experiences, show a marked reduction in their ability to interoceptively adapt through the anterior insula's top-down prediction of future states.

We have proposed a neuroanatomical processing model as a heuristic guide to understand how one can link optimal performance to how the individual "feels inside". This model focuses on the notion of a body prediction error, such as the difference between the value of the anticipated/predicted state and the value of the current interoceptive state, and consists of four components. First, information from peripheral receptors ascends via two different pathways, the A-B-fiber discriminative pathway that conveys precise information about the "what" and "where" of the stimulus impinging on the body, and the C-fiber pathway that conveys spatially- and time-integrated affective information (Craig 2007). These afferents converge via several waystations to the sensory cortex and the posterior insular cortex to provide a sense of the current body state. Second, centrally generated interoceptive states, for example via contextual associations from memory, reach the insular cortex via temporal and parietal cortex to generate body states based on conditioned associations (Gray and Critchley 2007, Yaguez et al. 2005). Third, within the insular cortex there is a dorsal-posterior to inferior-anterior organization from granular to agranular, which provides an increasingly "contextualized" representation of the interoceptive state (Shipp 2005), irrespective of whether it is generated internally or via the periphery. These interoceptive states are made available to the orbitofrontal cortex for context-dependent valuation (Rolls 2004, Kringelbach 2005) and to the anterior cingulate cortex for error processing (Critchley et al. 2005, Carter et al. 1998) and action valuation (Rushworth and Behrens 2008; Goldstein et al. 2007). Fourth, bidirectional connections to the basolateral amygdala (Augustine 1985, Jasmin, et al. 2004, Reynolds and Zahm 2005) and the striatum (Chikama et al. 1997), particularly ventral striatum (Fudge et al. 2005), provide the circuitry to calculate a body prediction error (similar to reward prediction error (Pessiglione et al. 2006, Preuschoff, Quartz, and Bossaerts 2008, Schultz and Dickinson 2000), and provide a neural signal for salience and learning. The insular cortex relays information to other brain systems to initiate motivated action to achieve a steady state (Craig 2007) by minimizing the body state prediction error. Thus the insular cortex is centrally located within a network of structures that are important for modulating processing according to internal and external demands.

The neuroscience approach to understanding optimal performance in extreme environments has several advantages over traditional descriptive approaches. First, once the role of specific neural substrates is identified, they can be targeted for interventions. Second, studies of specific neural substrates involved in performance in extreme environments can be used to determine what cognitive and affective processes are important for modulating optimal performance. Third, quantitative assessment of the contribution of different neural systems to performance in extreme environments could be used as indicators of training status or preparedness. The observation that the insular cortex and amygdala are associated with levels of resilience is a first step in bringing neuroscience approaches to a better understanding of what makes individuals perform differently when exposed to extreme environments. The application of this systems neuroscience approach will help to extend findings from specific studies with individuals exposed to extreme environments to develop a more general theory. As a consequence, one can begin to develop a rational approach to develop strategies to improve performance in these environments.

References

Arce, E., Simmons, A.N., Stein, M.B., Winkielman, P., Hitchcock, C. and Paulus, M.P. (2009) "Association between Individual Differences in Self-Reported Emotional Resilience and the Affective Perception of Neutral Faces", *J Affect Disord*, 114, 286–93, epub (2008).

The American Psychiatric Association (1994) *Diagnostic and Statistical Manual of Mental Disorders, 4th Ed. (DSM IV)*, (Washington, DC: The American Psychiatric Association).

Augustine, J.R. (1985) "The Insular Lobe in Primates Including Humans", *Neurol Res*, 7, 2–10.

Augustine, J.R. (1996) "Circuitry and Functional Aspects of the Insular Lobe in Primates Including Humans", *Brain Res Brain Res Rev*, 22, 229–44.

Banzett, R.B., Mulnier, H.E., Murphy, K., Rosen, S.D., Wise, R.J. and Adams, L. (2000) "Breathlessness in Humans Activates Insular Cortex", *Neuroreport*, 11, 2117–20.

Bonanno, G.A. (2004) "Loss, Trauma, and Human Resilience: Have We Underestimated the Human Capacity to Thrive after Extremely Aversive Events?" *Am Psychol*, 59, 20–28.

Boynton, G.M., Engel, S.A., Glover, G.H. and Heeger, D.J. (1996) "Linear Systems Analysis of Functional Magnetic Resonance Imaging in Human V1", *J Neurosci*, 16, 4207–21.

Campbell-Sills, L., Cohan, S.L. and Stein, M.B. (2006) "Relationship of Resilience to Personality, Coping, and Psychiatric Symptoms in Young Adults", *Behav Res Ther*, 44, 585–99.

Carter, C.S., Braver, T.S., Barch, D.M., Botvinick, M.M., Noll, D. and Cohen, J.D. (1998) "Anterior Cingulate Cortex, Error Detection, and the Online Monitoring of Performance", *Science*, 280, 747–9.

Charney, D. (2006) "The Psychobiology of Resilience to Extreme Stress: Implications for the Prevention and Treatment of Mood and Anxiety Disorders", *Biological Psychiatry*, 59, 91S.

Chikama, M., McFarland, N.R., Amaral, D.G. and Haber, S.N. (1997) "Insular Cortical Projections to Functional Regions of the Striatum Correlate with Cortical Cytoarchitectonic Organization in the Primate", *J Neurosci*, 17, 9686–9705.

Cohen, M.S. (1997) "Parametric Analysis of fMRI Data Using Linear Systems Methods", *Neuroimage*, 6, 93–103.

Connor, K.M. and Davidson, J.R. (2003) "Development of a New Resilience Scale: The Connor-Davidson Resilience Scale (CD-RISC)", *Depress Anxiety*, 18, 76–82.

Cox, R.W. (1996) "AFNI: Software for Analysis and Visualization of Functional Magnetic Resonance Neuroimages", *Comput Biomed Res*, 29, 162–73.

Craig, A.D. (2002) "How Do You Feel? Interoception: The Sense of the Physiological Condition of the Body", *Nat Rev Neurosci*, 3, 655–66.

Craig, A.D. (2003a) "A New View of Pain as a Homeostatic Emotion", *Trends Neurosci*, 26, 303–7.

Craig, A.D. (2003b) "Interoception: The Sense of the Physiological Condition of the Body", Curr *Opin Neurobiol*, 13, 500–505.

Craig, A.D. (2007) "Interoception and Emotion: A Neuroanatomical Perspective" in Handbook of Emotions (3d Ed.), Lewis, M., Haviland-Jones, J. and Feldman Barrett, L. (eds.), (New York City, NY: Guilford Press), 272–90.

Craig, A.D. (2009) "How Do You Feel—Now? The Anterior Insula and Human Awareness", *Nat Rev Neurosci*, 10, 59–70.

Craig, A.D. and Bushnell, M.C. (1994) "The Thermal Grill Illusion: Unmasking the Burn of Cold Pain", *Science*, 265, 252–5.

Critchley, H.D., Tang, J., Glaser, D., Butterworth, B. and Dolan, R.J. (2005) "Anterior Cingulate Activity During Error and Autonomic Response", *Neuroimage*, 27, 885–95.

Critchley, H.D., Wiens, S., Rotshtein, P., Ohman, A. and Dolan, R.J. (2004) "Neural Systems Supporting Interoceptive Awareness", *Nat Neurosci*, 7, 189–95.

Damasio, A.R. (1994) "Descartes' Error and the Future of Human Life", *Sci Am*, 271, 144.

Davidson, J.R., Payne, V.M., Connor, K.M., Foa, E.B., Rothbaum, B.O., Hertzberg, M.A. and Weisler, R.H. (2005) "Trauma, Resilience and Saliostasis: Effects of Treatment in Post-Traumatic Stress Disorder", *Int Clin Psychopharmacol*, 20, 43–8.

Eddy, W.F., Fitzgerald, M. and Noll, D.C. (1996) "Improved Image Registration by Using Fourier Interpolation", *Magn Reson Med*, 36, 923–31.

Eisenberger, N.I., Lieberman, M.D. and Williams, K.D. (2003) "Does Rejection Hurt? An fMRI Study of Social Exclusion", *Science*, 302, 290–92.

Feinle, C. (1998) "Role of Intestinal Chemoreception in the Induction of Gastrointestinal Sensations", *Dtsch Tierarztl Wochenschr*, 105, 441–4.

Fist, M., Spitzer, R., Gibbon, M. and Williams, J. (1995) "Structured Clinical Interview for DSM IV Axis I Disorders—Patient Edition", (New York City, NY: New York State Psychiatric Institute, Biometrics Research Department) SCID-I/P, version 2.0.

Forman, S.D., Cohen, J.D., Fitzgerald, M., Eddy, W.F., Mintun, M.A. and Noll, D.C. (1995) "Improved Assessment of Significant Activation in Functional Magnetic Resonance Imaging (fMRI): Use of a Cluster-Size Threshold", *Magn Reson Med*, 33, 636–47.

Fredrickson, B.L. (2004) "The Broaden-and-Build Theory of Positive Emotions", *Philos Trans R Soc Lond B Biol Sci*, 359, 1367–78.

Friston, K.J., Frith, C.D., Turner, R. and Frackowiak, R.S. (1995) "Characterizing Evoked Hemodynamics with fMRI", *Neuroimage*, 2, 157–65.

Fudge, J.L., Breitbart, M.A., Danish, M. and Pannoni, V. (2005) "Insular and Gustatory Inputs to the Caudal Ventral Striatum in Primates", *J Comp Neurol*, 490, 101–18.

Garavan, H., Ross, T.J. and Stein, E.A. (1999) "Right Hemispheric Dominance of Inhibitory Control: An Event-Related Functional MRI Study", *Proc Natl Acad Sci USA*, 96, 8301–6.

Goldstein, R.Z., Tomasi, D., Rajaram, S., Cottone, L.A., Zhang, L., Maloney, T., Telang, F., Alia-Klein, N. and Volkow, N.D. (2007) "Role of the Anterior Cingulate and Medial Orbitofrontal Cortex in Processing Drug Cues in Cocaine Addiction", *Neuroscience*, 144, 1153–9.

Gray, M.A. and Critchley, H.D. (2007) "Interoceptive Basis to Craving", *Neuron*, 54, 183–6.

Hariri, A.R., Drabant, E.M., Munoz, K.E., Kolachana, B.S., Mattay, V.S., Egan, M.F. and Weinberger, D.R. (2005) "A Susceptibility Gene for Affective Disorders and the Response of the Human Amygdala", *Arch Gen Psychiatry*, 62, 146–52.

Hariri, A.R., Mattay, V.S., Tessitore, A., Kolachana, B., Fera, F., Goldman, D., Egan, M.F. and Weinberger, D.R. (2002) "Serotonin Transporter Genetic Variation and the Response of the Human Amygdala", *Science*, 297, 400–403.

Haxby, J.V., Hoffman, E.A. and Gobbini, M.I. (2000) "The Distributed Human Neural System for Face Perception", *Trends Cogn Sci*, 4, 223–33.

Huettel, S.A., Misiurek, J., Jurkowski, A.J. and McCarthy, G. (2004) "Dynamic and Strategic Aspects of Executive Processing", *Brain Res*, 1000, 78–84.

Jasmin, L., Burkey, A.R., Granato, A. and Ohara, P.T. (2004) "Rostral Agranular Insular Cortex and Pain Areas of the Central Nervous System: A Tract-Tracing Study in the Rat", *J Comp Neurol*, 468, 425–40.

Johnstone, T., Somerville, L.H., Alexander, A.L., Oakes, T.R., Davidson, R.J., Kalin, N.H. and Whalen, P.J. (2005) "Stability of Amygdala BOLD Response to Fearful Faces over Multiple Scan Sessions", *Neuroimage*, 25, 1112–23.

Kringelbach, M.L. (2005) "The Human Orbitofrontal Cortex: Linking Reward to Hedonic Experience", *Nat Rev Neurosci*, 6, 691–702.

Lahuerta, J., Bowsher, D., Campbell, J. and Lipton, S. (1990) "Clinical and Instrumental Evaluation of Sensory Function before and after Percutaneous Anterolateral Cordotomy at Cervical Level in Man", *Pain*, 42, 23–30.

LaMotte, R.H., Thalhammer, J.G., Torebjork, H.E. and Robinson, C.J. (1982) "Peripheral Neural Mechanisms of Cutaneous Hyperalgesia Following Mild Injury by Heat", *J Neurosci*, 2, 765–781.

Light, A.R. and Perl, E.R. (2003) "Unmyelinated Afferent Fibers Are Not Only for Pain Anymore", *J Comp Neurol*, 461, 137–9.

Luthar, S.S., Cicchetti, D. and Becker, B. (2000) "The Construct of Resilience: A Critical Evaluation and Guidelines for Future Work", *Child Dev*, 71, 543–62.

Mesulam, M.M. and Mufson, E.J. (1982) "Insula of the Old World Monkey. III: Efferent Cortical Output and Comments on Function", *J Comp Neurol*, 212, 38–52.

Nimchinsky, E.A., Gilissen, E., Allman, J.M., Perl, D.P., Erwin, J.M. and Hof, P.R. (1999) "A Neuronal Morphologic Type Unique to Humans and Great Apes", *Proc Natl Acad Sci USA*, 96, 5268–73.

Norusis, M. (1990) "SPSS Base System User's Guide", (Chicago, IL: SPSS, Inc.).

Olausson, H., Lamarre, Y., Backlund, H., Morin, C., Wallin, B.G., Starck, G., Ekholm, S., Strigo, I., Worsley, K., Vallbo, A.B. and Bushnell, M.C. (2002) "Unmyelinated Tactile Afferents Signal Touch and Project to Insular Cortex", *Nat Neurosci*, 5, 900–904.

Paulus, M.P. (2007) "Decision-Making Dysfunctions in Psychiatry-Altered Homeostatic Processing?" *Science*, 318, 602–6.

Paulus, M.P., Feinstein, J.S., Castillo, G., Simmons, A.N. and Stein, M.B. (2005) "Dose-Dependent Decrease of Activation in Bilateral Amygdala and Insula by Lorazepam During Emotion Processing", *Arch Gen Psychiatry*, 62, 282–8.

Paulus, M.P., Potterat, E., Taylor, M., Van Orden, K., Bauman, J, Momen, N., Padilla, G. and Swain, J. (2009, in press) "A Neuroscience Approach to Optimizing Brain Resources for Human Performance in Extreme Environments", *Neurosci Biobehav Res*.

Paulus, M.P., Rogalsky, C., Simmons, A., Feinstein, J.S. and Stein, M.B. (2003) "Increased Activation in the Right Insula During Risk-Taking Decision Making is Related to Harm Avoidance and Neuroticism", *Neuroimage*, 19, 1439–48.

Paulus, M.P. and Stein, M.B. (2006) "An Insular View of Anxiety", *Biol Psychiatry*, 60, 383–7.

Pessiglione, M., Seymour, B., Flandin, G., Dolan, R.J. and Frith, C.D. (2006) "Dopamine-Dependent Prediction Errors Underpin Reward-Seeking Behaviour in Humans", *Nature*, 442, 1042–5.

Phan, K.L., Wager, T., Taylor, S.F. and Liberzon, I. (2002) "Functional Neuroanatomy of Emotion: A Meta-Analysis of Emotion Activation Studies in PET and fMRI", *Neuroimage*, 16, 331–48.

Preuschoff, K., Quartz, S.R. and Bossaerts, P. (2008) "Human Insula Activation Reflects Risk Prediction Errors as Well as Risk", *J Neurosci*, 28, 2745–52.

Reynolds, S.M. and Zahm, D.S. (2005) "Specificity in the Projections of Prefrontal and Insular Cortex to Ventral Striatopallidum and the Extended Amygdala", *J Neurosci, 25,* 11757–67.

Richardson, G.E. (2002) "The Metatheory of Resilience and Resiliency", *J Clin Psychol*, 58, 307–21.

Robinson, S.K., Viirre, E.S., Bailey, K.A., Gerke, M.A., Harris, J.P. and Stein, M.B. (2005) "Randomized Placebo-Controlled Trial of a Selective Serotonin Reuptake Inhibitor in the Treatment of Nondepressed Tinnitus Subjects", *Psychosom Med*, 67, 981–8.

Rolls, E.T. (2004) "The Functions of the Orbitofrontal Cortex", *Brain Cogn*, 55, 11–29.

Rushworth, M.F. and Behrens, T.E. (2008) "Choice, Uncertainty and Value in Prefrontal and Cingulate Cortex", *Nat Neurosci*, 11, 389–97.

Schmelz, M., Schmidt, R., Bickel, A., Handwerker, H.O. and Torebjork, H.E. (1997) "Specific C-Receptors for Itch in Human Skin", *J Neurosci*, 17, 8003–8.

Schultz, W., Dayan, P. and Montague, P.R. (1997) "A Neural Substrate of Prediction and Reward", *Science*, 275, 1593–9.

Schultz, W. and Dickinson, A. (2000) "Neuronal Coding of Prediction Errors", *Annu Rev Neurosci*, 23, 473–500.

Shipp, S. (2005) "The Importance of Being Agranular: A Comparative Account of Visual and Motor Cortex", *Philos Trans R Soc Lond B Biol Sci*, 360, 797–814.

Simmons, A., Matthews, S.C., Paulus, M.P. and Stein, M.B. (2008) "Intolerance of Uncertainty Correlates with Insula Activation During Affective Ambiguity", *Neurosci Lett*, 430, 92–7.

Simmons, A., Matthews, S.C., Stein, M.B. and Paulus, M.P. (2004) "Anticipation of Emotionally Aversive Visual Stimuli Activates Right Insula", *Neuroreport*, 15, 2261–5.

Simmons, A., Strigo, I., Matthews, S.C., Paulus, M.P. and Stein, M.B. (2006) "Anticipation of Aversive Visual Stimuli is Associated with Increased Insula Activation in Anxiety-Prone Subjects", *Biol Psychiatry*, 60, 402–9.

Simmons, A., Strigo, I.A., Matthews, S.C., Paulus, M.P. and Stein, M.B. (2009) "Initial Evidence of a Failure to Activate Right Anterior Insula During Affective Set Shifting in Posttraumatic Stress Disorder", *Psychosom Med*, 71, 373–7.

Spielberger, C. (1983) *Manual for the State-Trait Anxiety Inventory (Form Y)*, (Paolo, Alto, CA: Consulting Psychologists Press).

Stein, M.B., Simmons, A.N., Feinstein, J.S. and Paulus, M.P. (2007) "Increased Amygdala and Insula Activation During Emotion Processing in Anxiety-Prone Subjects", *Am J Psychiatry*, 164, 318–27.

Tracey, I., Becerra, L., Chang, I., Breiter, H., Jenkins, L., Borsook, D. and Gonzalez, R.G. (2000) "Noxious Hot and Cold Stimulation Produce Common Patterns of Brain Activation in Humans: A Functional Magnetic Resonance Imaging Study", *Neurosci Lett*, 288, 159–62.

Tugade, M.M., Fredrickson, B.L. and Barrett, L.F. (2004) "Psychological Resilience and Positive Emotional Granularity: Examining the Benefits of Positive Emotions on Coping and Health", *J Pers*, 72, 1161–90.

Vallbo, A.B., Olausson, H., Wessberg, J. and Kakuda, N. (1995) "Receptive Field Characteristics of Tactile Units with Myelinated Afferents in Hairy Skin of Human Subjects", *J Physiol*, 483 (Pt 3), 783–95.

Werner, E. and Smith, R. (1992) *Overcoming the Odds: High Risk Children from Birth to Adulthood*, (Ithaca, NY: Cornell).

Yaguez, L., Coen, S., Gregory, L.J., Amaro, E., Jr., Altman, C., Brammer, M.J., Bullmore, E.T., Williams, S.C. and Aziz, Q. (2005) "Brain Response to Visceral Aversive Conditioning: A Functional Magnetic Resonance Imaging Study", *Gastroenterology*, 128, 1819–29.

Chapter 11
The Cognitive Neuroscience of Insight and its Antecedents

John Kounios

Mark Beeman

On August 5, 1949, a fire burned out of control in Mann Gulch in Montana. A team of firefighters led by Wag Dodge parachuted into the gulch on the side opposite the fire. As the firefighters worked their way down the side of the gulch, the fire, whipped by fierce winds, suddenly jumped across the canyon and lit the grass below the men. The winds continued to propel the fire, which raced up the incline toward the firefighters. The men started making their way uphill away from the fire, but Dodge soon realized that the fire was approaching them too quickly for them to outrun it. In a flash of insight, he suddenly stopped running and lit the grass in front of him. This new fire burned the ground bare. He then lay down on the burnt ground and waited as the fire him approached him, surrounded him, and then passed him. Almost all the other firefighters died that day.

Dodge's solution to this problem was not new. The Indians of the Great Plains learned to use this technique long before Dodge did. But Dodge apparently never knew this. His solution was, for him, a remarkable and novel example of creative insight. He solved the problem of avoiding being burnt alive by doing the least obvious thing—setting a fire. And importantly, he was able to achieve this sudden insight in an emergency situation.

Most people are familiar with the phenomenon of *insight,* the sudden solution to a seemingly intractable problem or a sudden comprehension that allows one to see a situation in a new light. Such insights are powerful experiences because they come unexpectedly and because they reorganize one's thoughts and perceptions in novel ways. However, even though sudden insights seem to be disconnected from the ongoing stream of thought and appear to come from nowhere, a recent series of behavioral and neuroimaging studies have shown that an insight is actually the culmination of a series of brain states and processes operating at a number of time scales.

The Approach

Obviously, it isn't practical to put test subjects in a brain scanner and wait until they have an insight. Instead, the subjects' brain activity was measured while they solved a series of simple verbal problems that can be solved either by insight or by a conscious, methodical, strategy which cognitive psychologists refer to as *analytical*. The subjects were presented with a series of *compound remote associates* problems (Bowden and Beeman 2003), each of which consisted of three words (for example, *pine, sauce, crab*). The subject's task was to derive a fourth word that could form a compound or familiar phrase with each of the three words of the problem (that is, *apple-pineapple, applesauce, crabapple*). Each time a subject solved one of these problems, he or she was prompted to indicate whether the solution had popped into awareness as a sudden insight or whether the solution had been the result of methodical, conscious, hypothesis testing. Each subject's solutions were sorted out according to these solution methods and then were examined and compared with the corresponding patterns of brain activity.

The Neural Correlate of the "*Aha!* Moment"

The first neuroimaging study of insight (Jung-Beeman et al. 2004) included two separate experiments, one involving measuring neural activity with high-density electroencephalograms (EEG), the other measuring blood flow in the brain using functional magnetic resonance imaging (fMRI). These two techniques are complimentary, because EEG has excellent temporal resolution accompanied by modest spatial resolution, while fMRI has excellent spatial resolution but only modest temporal resolution. Together, these two techniques allowed us to localize insight in the brain in both space and time.

Comparing neural activity for insight and analytic solutions at the point of problem solution, fMRI revealed one statistically significant brain activation. There was greater brain activity for insight solutions in the anterior superior temporal gyrus of the right temporal lobe. This was confirmed by EEG, which found a burst of high-frequency (that is, gamma band) neural activity measured in electrodes over this location occurring at about the point in time at which the insight popped into awareness. Interestingly, the right anterior superior temporal gyrus is a brain area implicated in integrative conceptual processing, such as occurs in the processing of metaphors and jokes.

Sensory Gating

An additional finding turned up in the study of Jung-Beeman et al. (2004) that was a surprise. The EEG data showed a burst of insight-related alpha-band (approximately 10 Hz) activity at about 1.5 seconds before the button-press response indicating that a problem had been solved. This burst of alpha-band activity occurred just before the burst of gamma-band activity that coincided with the problem solution popping into awareness and was detected over right occipital cortex. Ever since the 1920s, posterior alpha has been associated with cortical idling or inhibition of the visual system. Specifically, alpha is inversely proportional to cortical activity and, in the case of posterior alpha, it represents gating or reduction of sensory inputs.

A Hemispheric Model of Insight

Although this finding of sensory gating was a surprise, it fits into the emerging model of insight problem solving. According to this view (Jung-Beeman et al. 2004), the cerebral hemispheres process semantic information differently from each other. Specifically, the left hemisphere performs *fine semantic coding* in which a concept activates, by association, a small number of closely related concepts. For example, the concept *table* might activate only *chair* and *lamp*. In contrast, the right hemisphere performs *coarse semantic coding* in which a concept weakly activates a larger number of more distantly associated concepts. In this case, for example, the concept *table* might activate *water* (such as water-table), *payment* (for "paying under the table"), *ping-pong* (such as ping-pong table), and so on. According to the model, an insight occurs when attention switches from closely associated, strongly-activated, left-hemisphere solution candidates to a weakly activated, distantly associated, right-hemisphere solution. The hypothesis was that the alpha burst reduced potentially interfering visual inputs to the right hemisphere, thereby facilitating the retrieval of a weakly activated solution. A particularly interesting aspect of this model is that, if this interpretation is correct, the burst of sensory-gating alpha reflects the brain's sensitivity or "awareness" of a subconscious solution represented in the right hemisphere.

Preparatory Effects

The presence of the alpha burst showed that, even though the phenomenal awareness of an insight is sudden, discrete, and unpredictable, the "*Aha!*" itself has a neural antecedent which presumably facilitates the insight. But the existence of this antecedent suggested the possibility that there were additional upstream antecedents. This notion was tested in a subsequent EEG/fMRI study (Kounios et al. 2006) using the same set of compound remote associate problems. This study examined neural activity during the two seconds immediately preceding the presentation of each problem. In the EEG study, when ready to view the next problem in the sequence, subjects made a bimanual button-press which initiated the display of the problem one second later. (For technical reasons, the inter-trial intervals were randomly varied and not subject-controlled in the fMRI study.)

Most of the results from the EEG and fMRI studies closely corresponded. During the pre-problem interval, neural activity was greater in a number of brain areas preceding problems that a subject would subsequently solve with insight (relative to those solved analytically). These areas included the anterior cingulate, which is closely associated with cognitive control mechanisms such as those involved in attention switching and detection of competing cognitive and response representations, and bilateral temporal-lobe activity consistent with the priming of left- and right-hemisphere areas involved in semantic processing. In contrast, results for the interval preceding problems subsequently solved analytically revealed increased activity in the visual system.

Together, these results suggest neural preparatory mechanisms that bias a person to solve an upcoming problem either with insight or analytically. Insight solving is facilitated by priming of the anterior cingulate, which is involved in attention switching and the detection of alternative solution candidates, and bilateral temporal areas associated with semantic processing. Analytical solving is associated with increased activity in the visual system, which may simply reflect greater attention to the monitor on which the problem will subsequently be displayed. Another way of thinking about this is that neural preparation for insight solving involves focusing attention inwardly on internal conceptual representations and on processes responsible for the cognitive flexibility necessary for insight while preparation for analytical processing involves focusing attention outwardly.

So far, we've traced the antecedents of insight backward in time from the gamma-band burst reflecting the insight itself to the immediately preceding alpha-band burst reflecting sensory gating, to the preparatory effects preceding the presentation of the problem. These antecedents constitute a chain of events that enable an insight to occur. But how far back in time can these antecedents be traced? What is the matrix from which this chain of events emerges? To investigate this question, the study looked as far back in time as it could—*resting-state brain activity*.

Resting-State Brain Activity

Cognitive activity can be roughly categorized into two types: *directed* and *undirected*. The vast majority of studies of human cognition, whether behavioral or neuroscience, examine the directed form of cognition in which a subject is given a specific task to perform. In contrast, studies of undirected cognition, though growing in number, are still in the minority. Undirected cognition is the spontaneous mental activity that a person engages in when there is no explicit task or goal. Colloquially put, it is daydreaming. In cognitive neuroscience, it is called *resting-state brain activity*.

Such resting-state brain activity has been a focus of clinical EEG research for many years (John et al. 1988) because of evidence that individual differences in resting-state activity reflect specific neurological and psychiatric disorders. More recently, fMRI studies of nonclinical subjects indicate that individual differences in resting-state activity correlate with differences in personality (Kumari et al. 2004). With this in mind, we investigated whether individual differences in resting-state EEG activity correlate with the tendency to solve problems with insight rather than analytically (Kounios et al. 2008).

In this study, the first step was to record a few minutes of eyes-closed resting-state activity from a group of subjects with a high-density EEG array (128-channels). These subjects did not know what they would be doing afterwards, so their brain activity could not reflect any specific expectation about the nature of the subsequent task. Then, they were given a series of anagrams to solve. (It was decided to use anagrams, rather than compound remote associates, in this study in order to start to generalize the results to other types of problems.) This phase used the same insight judgment procedure that was used in the previous studies with compound remote associates. On each trial, a subject viewed an anagram and when (and if) the subject derived the solution, she or he pressed a button immediately and was then prompted to verbalize the solution. Subsequently, they were prompted to indicate whether the solution had been derived with insight or analytically.

The subjects were then divided into two groups defined by whether they solved most of the anagrams with insight or solved most of them analytically. Then, the resting-state EEG activity of these groups was analyzed separately and compared in terms of EEG power and topography using the standard classical EEG frequency bands. Two general patterns of results were predicted, based on previous behavioral and neuroscience work on creativity.

First, prior research on creativity has shown that creative individuals tend to have tonically diffuse, rather than focused, attention (Ansburg and Hill 2003). This can be manifested as distractibility. However, such diffuse attention is thought to benefit creative individuals by allowing them to sample a greater range of stimuli in the environment. By taking in a greater range of information, this increases the likelihood that a stray stimulus will trigger a remote association that will lead to the solution of a problem. Therefore a greater diffuse activation of visual cortex was predicted, manifested as less alpha-band activity and less beta-band activity.

Second, diffuse *perceptual* attention is thought to be related to diffuse *conceptual* attention, which is defined as the tendency to think in terms of remote associations. And, as discussed above, thinking in terms of remote associations is theorized to be a function of right-hemisphere activity. Therefore, greater general right-hemisphere brain activity was predicted in high-insight subjects, particularly in frontal, temporal, and parietal areas associated with semantic information processing.

What was found was that the high-insight and high-analytical groups differed in resting-state activity in every EEG frequency band. The patterns of results were complicated, but supported the predictions. High-insight subjects showed less alpha- and beta-band activity measured over visual cortex. They also showed greater right-hemisphere activity in a number of brain areas.

These results show that resting-state brain activity—in the absence of specific expectations or goals—biases a person toward either insight or analytical processing when subsequently given problems to solve. This means that cognitive problem-solving strategies do not arise in a "vacuum." They are biased by preexisting patterns of brain activity which themselves are likely to result from subtle individual differences in neuroanatomy, cytoarchitectonics, or neurochemistry.

However, what these results do *not* reveal is whether the measured individual differences in insight-related resting-state activity are stable or transient. In general, individual differences in resting-state activity are known to be fairly stable over time and genetically influenced. However, it is not yet known whether the subset of resting-state activity that is insight related is also stable and genetically loaded.

Implications and Future Directions

Overall, this line of research shows that insight, while phenomenologically discrete, is actually the culmination of a series of brain states and processes extending over time and operating at a number of time scales ranging from milliseconds to (at least) hours. This suggests that there are a number of points in this cascade of processes that are available for external influence. Future research will investigate various possibilities for influencing these processes as a step toward developing a technology for systematically influencing cognitive style according to situational and task demands. These methods are likely to include contextual manipulation, cognitive training, pharmacology, and direct brain stimulation. It is an important goal to develop methods to maximize the likelihood that people can bring the optimal cognitive strategies to bear in a variety of critical situations so that more people react like Wag Dodge rather than like his less creative and less flexible colleagues.

References

Ansburg, P.I., & Hill, K. (2003) "Creative and Analytic Thinkers Differ in Their Use of Attentional Resources", *Personality and Individual Differences,* 34, 1141–52.

Bowden, E.M. and Jung-Beeman, M. (2003) "One Hundred and Forty-Four Compound Remote Associate Problems: Short Insight-Like Problems with One-Word Solutions", *Behavioral Research, Methods, Instruments, and Computers,* 35, 634–9.

Jung-Beeman, M., Bowden, E.M., Haberman, J., Frymiare, J.L., Arambel-Liu, S., Greenblatt, R., Reber, P.J. and Kounios, J. (2004) "Neural Activity when People Solve Verbal Problems with Insight", *PLoS Biology,* 2, 500–510.

John, E.R., Prichep, L.S., Fridman, J. and Easton, P. (1988) "Neurometrics: Computer-Assisted Differential Diagnosis of Brain Dysfunctions", *Science*, 239, 162–9.

Kounios, J., Fleck, J.I., Green, D.L., Payne, L., Stevenson, J.L., Bowden, M. and Jung-Beeman, M. (2008) "The Origins of Insight in Resting-State Brain Activity", *Neuropsychologia*, 46, 281–91.

Kounios, J., Frymiare, J.L., Bowden, E.M., Fleck, J.I., Subramaniam, K., Parrish, T.B., Jung-Beeman, M. (2006). "The Prepared Mind: Neural Activity Prior to Problem Presentation Predicts Subsequent Solution by Sudden Insight", *Psychological Science*, 17, 882–90.

Kumari, V., Fytche, D.H., Williams, S.C.R. and Gray, J. (2004) "Personality Predicts Brain Responses to Cognitive Demands", *Journal of Neuroscience*, 24, 10636–41.

Chapter 12
Decision-making Under Risk and Stress: Developing a Testable Model

Richard Gonzalez

Israel Liberzon

Decisions are made throughout our daily life, even though most of the time we are unaware that we make many of those decisions. Changing lanes on the highway, choosing to check email, deciding what to make for dinner, deciding to take aspirin to alleviate a headache are all examples of those daily decisions that seem not to garner much awareness that a decision is involved. The more salient decisions however are those that often involve stress of some kind: a time deadline, a decision with major financial implications, health, or one in which we experience a great amount of subjective uncertainty about a tradeoff. These types of stress though are related to the process of decision-making itself. That is, how we go about deciding (for example, the tradeoffs we make) and the constraints of our decision (including time constraints and budget constraints) can make decision-making subjectively stressful.

Another type of stress arises due to the situation itself, independently of the decision-making that is supposed to take place. This type of stress creates a context in which the effects of stress on decision-making may be different than the effects of stress as defined by time pressure or difficult tradeoffs inherent to decision-making itself. This kind of stress is less specific to decision-making and it is has been studied, modeled and characterized in psychological, physiological and clinical literature (McEwen 2000, de Kloet 2005). Understanding the relation between this type of stress and decision-making can open a new window into a completely new area previously unaddressed by decision-making research. It can help us to understand the psychological and physiological processes that govern compromised or impaired decision-making that often accompanies conditions of severe or pathological stress like bereavement, depression or post-traumatic stress disorder. This brief report addresses relevant research at the intersection of stress and decision-making with the goal of outlining a preliminary model about the underlying neurocircuitry.

Thus while the stress literature has described different components (or types) of stress reactions and there is a strong tradition of research on behavioral decision-making (Broadbent 1971, and Janis and Mann 1977), the study at the intersection of stress and decision-making remains fairly simple in its analysis.

How do different types of stress influence different types of decisions? Does our current understanding of neuroanatomy and neurophysiology provide useful tools in understanding the intersection of stress and decision-making? This brief review carves out a piece of the intersection of stress and decision-making and provides a model for the role neuroscience can play in understanding the interplay. We first review some relevant decision-making models and then review how stress might be involved in decision-making and how it may be incorporated in decision-making models.

Behavioral Decision-making Models

The field of decision-making focuses on three major types of decisions: 1) multiattribute choice, 2) decision-making under risk, and 3) decision-making under uncertainty.

1. Multiattribute choice involves deciding between options where each option has many dimensions. For example, when purchasing a digital camera one may consider price, megapixels, zoom, and storage capacity. Each camera has a value on each of those dimensions and the decision maker's problem is to choose the camera that provides a best fit given the requirements (for example, a camera under $800 that can take pictures of your daughter playing soccer).
2. Decision-making under risk involves deciding between options that offer probabilistic outcomes. For example, the oncologist tells the patient the probabilities of successful outcomes associated with a menu of treatment options and the patient uses those probabilities to decide on a particular course of treatment. In this case the probability is "known" to the decision maker because it was given by the oncologist. Such a known probability is what is meant by "risk" in the decision-making literature. The term risk applies to the case where probabilities are given explicitly and are not necessarily associated with a negative outcome. For example, decision-making under risk applies to contexts even when there are no losses (for example, a choice between receiving $50 for sure and playing a gamble that yields $100 at 50 percent and $10 at 50 percent falls in the domain of decision-making under risk).
3. The third type is called decision-making under uncertainty and refers to cases where the probability is not known as a numerical value. Decision-making under uncertainty includes deciding between options where the probability of the event is described verbally ("highly likely", "more likely than not", "somewhat probable", and so on) or the decision options are described as conditional propositions ("if it rains tomorrow we will do X, otherwise we will do Y", "if the diagnostic test is positive, then we will follow this treatment plan").

To better understand processes involved in decision-making and to be able to predict behavioral choices traditionally, formal mathematical models that accurately predict decision-making outcomes under various conditions had been used. To this end, the underlying mathematical models for the three types of decisions listed above are different: multiattribute choice does not involve probability; decision-making under risk involves probability; and decision-making under uncertainty involves ambiguity about chances or a lack of numerical precision about the associated probabilities. A common element of many such decision-making models involves a weighted representation of some kind. A simple multiattribute choice model defines subjective values on each dimension (such as how much the decider values an extra megabyte in computer RAM or an extra year of life under a particular level of quality of life, values that may not be linear in their variable), weights each of those subjective values with subjective weights that index the importance of each dimension, and takes a weighted sum to evaluate each option. A simple multiattribute choice representation is $\Sigma w_i v_i(x_i)$ for dimension *i* that has an importance weight w and a transformation v of the original dimension value. The choice mechanism then selects the option with the greatest weighted sum. A classic reference for this model is Keeny (1992).

A standard model for decision-making under risk is prospect theory, which also follows a weighted representation of a special form. A probabilistic prospect is represented as $\Sigma[w(p_{1-i)}-w(p_{1-(i-1)})]v(x_i)$. This complicated form allows for proper summation with nonlinear weights on the probabilities. The logic of this representation is that each outcome is transformed to its subjective value, weighted by the marginal cumulative transformed weight, and summed across outcomes. The option with the greatest cumulative weighted sum is selected. Particular functional forms for weighting function *w* and value function *v* have been proposed, but for our purposes we focus on the qualitative aspects of *w* and *v*. Behavioral data suggest that in the case of probabilities that are explicitly given, people place more weight on small probabilities and less weight on large probabilities in a manner that leads to an inverse S-shape transformation (concave for small *p*, convex for large *p*). Further, behavioral data suggests that value is defined with respect to changes from a reference point, and thus the value function *v* is defined with respect to gains and losses. This is perhaps the most important addition prospect theory made to the generalized expected utility framework. An important behavioral observation is that in many domains, losses receive more weight than gains. These two properties of *w* and *v* can account for two broad classes of behavioral decision-making results. The weighting function *w* can account for observations of risk aversion and risk seeking; the value function *v* can account for the behavioral observation that "losses loom larger than gains." Two classic references in this behavioral decision-making tradition are Kahneman and Tversky (1979) and Tverksy and Kahneman (1992), research that lead to a Nobel prize in Economics.

The third type of decision-making model, decision-making under uncertainty, has received relatively little attention. This form of decision-making model also follows a weighted sum representation, but is even more complicated than the representation outlined above for decision-making under risk because it involves a further mapping from the characterization of an event into its subjective probability. For references in the behavioral decision-making tradition see Fox and Tversky (1998) and Wu and Gonzalez (1999).

A major limitation of the behavioral decision-making tradition is that it tends to be descriptive in the sense that mathematical models are developed to account for properties of the data. In some cases the mathematical model is predictive of new empirical results and in this sense the research program has been successful (Wu and Gonzalez 1996), but a justifiable critique of this literature is that it is not constrained nor is it based on fundamental psychological or neuroscience principles that govern psychological and brain processes that are at the base of the decision-making process. This is most clearly seen in the partition of decision-making problems into three types (as reviewed above—multiattribute choice, decision-making under risk and decision-making under uncertainty). This partition follows not from psychological principles but mostly because of mathematical convenience in the development of the decision representation (different types of weighted sums depending on whether the problem involves probability, does not involve probability or involves uncertainty). This modeling bias makes it very difficult, however, to develop a better understanding of the various psychological or physiological variables that may be influencing the decision-making process under various circumstances.

Future progress in the field of decision-making is dependent therefore on the ability to decompose the process of making a decision into testable cognitive and affective components mediated by the dedicated underlying neurocircuitry. This necessarily will have to incorporate vast literature in cognitive science, social psychology, cognitive and affective neuroscience. This next step is especially important for the development of models for decision-making and stress, as it will have to link up large bodies of existing research—critical for future progress. It is unlikely for example that stress is implicated differently in a decision-making problem involving choice of nonprobabilistic outcomes, probabilistic outcomes or uncertain outcomes, the three major partitions of the decision-making literature. A deeper understanding of the role of stress in decision-making will emerge when we reconceptualize decision-making in terms of its cognitive and affective components, and when we provide a psychological, physiological and neuroanatomical foundation for behavioral concepts such as risk aversion and loss aversion.

In the last decade, a new and exciting program of research under the domain of neuroeconomics involving neuroimaging of decision-making processes has emerged, suggesting that mapping of some of the relevant mathematical/behavioral variables from the models described above onto the neuroanatomical substrate might be possible. Early studies provided some promising initial evidence that the behavioral concepts of the weighting and value functions might have meaningful neural correlates. The transformation of probabilities characterized by the weighting function have been associated with neural activation in the dorsolateral prefrontal cortex (Tobler et al. 2008), the ventral striatum (Hsu et al. 2009) and dorsal anterior cingulate cortex (Paulus et al. 2006). The behavioral concept of loss aversion has been linked to activity in the ventromedial prefrontal cortex and ventral striatum (Tom et al. 2007). These early studies use different paradigms of decision-making and it is not surprising that the findings reported in one study are not necessarily replicated in the others. However, they do provide critical data for initial neuroanatomical models of decision-making and further work is obviously required to replicate, reconcile and confirm these initial findings. This work is clearly in its early stages and much more research is needed to establish the functional neurocircuitry of decision-making and to understand the implications and boundary conditions.

Intersection of Decision-making and Stress

Much of the research in the behavioral decision-making tradition focuses on the subjective experience of stress. In experimental research this has been typically operationalized as a manipulation of time constraints, or of performance demands (Broadbent 1971). Stress in decision-making can arise because of external factors such as time constraints placed by the demands of the decision, or internal factors such as a decision involving a difficult tradeoff between several important dimensions (for example, some end-of-life decisions). Here the stressor is internal because the tradeoff itself is experienced as "stressful" or that the decision itself produces "subjective distress." These kinds of external and internal forms of stress undoubtedly influence decision-making, as many classic examples from the psychological decision-making literature show (Broadbent 1971, and Janis and Mann 1977). The kinds of variables that characterize much of this literature though focus on more behavioral aspects of distress and behavioral aspects of decisions; however, these kinds of stress conditions may play out differently than the environmental/physiological stress or traumatic stress that is antecedent to post-traumatic stress disorder (PTSD). Understanding the implications of severe situational, physiologic or traumatic stress on decision-making is critical however, if one aims to understand decision-making under extreme conditions or the interplay between stress-related psychopathology and decision-making.

To begin addressing these questions it is important to focus on stress concepts that are concretely defined in a way that have a biological component, for example, stress that activates the limbic-hypothalamo-pituitary-adrenal (LHPA) axis (Shin and Liberzon *in press*). Until recently there has not been much research attention given to forms of stress that activate the LHPA axis in the context of decision-making. Two recent studies examined decision-making under risk using a Trier Social Stress Test, which is designed to elicit robust LHPA response through an anticipated public speaking task. Preston et al. (2007) used the Iowa Gambling Task and Starcke et al. (2008) used the Game of Dice Task (GDT) as their respective decision-making tasks.

Both studies yielded similar results, suggesting that anticipatory stress leads to more disadvantageous results, except that Preston et al. found an interaction with gender but Starcke et al. did not, despite different decision-making tasks. These early studies are important because they manipulated stress level and measured physiological correlates of it, however they did not attempt to delineate effects of stress on formal behavioral parameters or the psychological process involved in decision-making. In line with our earlier recommendation of decomposing decision processes into a set of cognitive and affective processes, Starcke et al (2008) provide evidence that the stress manipulation did not influence the executive functioning part of the GDT, which provides a useful boundary condition in our understanding of the role of stress on decision-making.

One interesting feature of both the Preston et al. (2007) and Starcke et al. (2008) studies is that they examined the role of learning probabilities over repeated trials and the role of feedback. Much of the behavioral decision-making literature has focused exclusively on the actual decision under cases of known probability (risk), and not on learning and feedback over repeated trials. It is possible that stress influences such processes in very deep ways, and a complete understanding of the role of stress on decision-making should examine other variables such as learning. Further, both Preston et al. (2007) and Starcke et al. (2008) manipulated stress in manner that has been previously shown to activate the LHPA axis (Trier task), and Starcke et al. also included a manipulation check using salivary cortisol. A slightly different paradigm was used by Porcelli and Delgado (2009), who demonstrate that stress as manipulated through a cold pressor task (and verified with skin conductance) can exacerbate the reflection effect commonly observed in the behavioral literature (risk aversion in gain-only gambles and risk seeking in loss-only gambles). However, to our knowledge there does not exist to date published research that manipulates cortisol directly.

Recently our research group developed a paradigm involving an oral administration of hydrocortisone (synthetic cortisol analogue) during a gambling task to begin addressing the direct effects of elevated stress hormones on decision-making processes. Using a gambling paradigm that manipulates both the value and probabilities of the options in an orthogonal design and using gain, losses and mixed gambles, we find behavioral evidence for the usual weighting function and value function properties of probability transformation and loss aversion, respectively. These properties are interestingly moderated by a single dose hydrocortisone administration (confirmed by saliva cortisol levels) that is given in a double-blind placebo control fashion, and is not detectable by the subjects beyond chance level. Furthermore, our paradigm allows us to begin outlining the neurocircuitry underlying different parameters of decision-making process, described by the mathematical models, such as loss aversion, probability assessment and weighing function in the same subjects and the same conditions, allowing a clearer description of the neurocircuitry involved. Finally, the hydrocortisone administration modulated the activity in the same brain region associated with the parameters listed above in a manner that is entirely consistent with the behavioral effects observed, allowing us to test specific mediation models.

Our preliminary findings in concert with the reported results in the literature using the Trier task suggest that cortisol sensitizes reward processing and at the same time appears to blunt the processing of probability. These initial conclusions of course, require much more research before they are planted firmly in the literature. However, we view them as providing an extremely promising direction for a better understanding of the role of stress on decision-making, in a paradigm linking the underlying cognitive and affective processing with associated neural circuitry. The results suggest that despite our critique that the behavioral decision-making literature has not been founded on first principles, there might be evidence that two major properties identified by the behavioral work, the weighting function *w* and the value function *v*, might be subserved by two separate neuroanatomical systems. The study of the role of stress on decision-making promises to be an active area of research that can provide many deep clues to decision-making in health, under extreme stress and during disease process.

References

Broadbent, D.E. (1971) *Decision and Stress* (London: Academic Press).

de Kloet, E.R., Joels, M., Holsboer, F. (2005) "Stress and the brain: From adaptation to disease", *Nature Reviews Neuroscience:* 6(6), 463–75.

Fox, C. and Tversky, A. (1998) "A belief-based account of decision under uncertainty", *Management Science,* 44, 889–95.

Hsu, M., Krajbich, I., Zhao, C., and Camerer, C. (2009) "Neural response to reward anticipation under risk is nonlinear in probabilities", *Journal of Neuroscience,* 29, 2231–37.

Kahneman, D. and Tversky, A. (1979) "Prospect theory: An analysis of decision under risk", *Econometrica*, 47, 263–92.

Janis, I.L. and Mann, L. (1977). *Decision-making: A Psychological Analysis of Conflict, Choice, and Commitment* (New York: Free Press).

McEwen, B.S. (2000) "The neurobiology of stress: from serendipity to clinical relevance", *Brain Research,* 886(1–2) Special Issue, 172–89.

Paulus, M. and Frank, L. (2006) "Anterior cingulated activity modulates nonlinear decision weight function of uncertain prospects", *NeuroImage,* 30, 668–77.

Porcelli, A. and Delgado, M. (2009) "Acute stress modulates risk taking in financial decision-making", *Psychological Science,* 20, 278–83.

Preston, S., Buchanan, T., Stansfield, R. and Bechara, A. (2007) "Effects of anticipatory stress on decision-making in a gambling task", *Behavioral Neuroscience*, 121, 257–63.

Shin, L. and Liberzon, I. (*in press*) "The neurocircuitry of fear, stress, and anxiety disorders", *Neuropsychopharmacology Reviews.*

Starcke, K., Wolf, O., Markowitsch, H., and Brand, M. (2008) "Anticipatory stress influences decision-making under explicit risk conditions", *Behavioral Neuroscience,* 122, 1352–60.

Tobler, P., Christopoulos, G., O'Doherty, J., Dolan, R., and Schultz, W. "Neuronal distortions of reward probability without choice", *Journal of Neuroscience,* 28, 11703–11711.

Tom, S., Fox, C., Trepel, C., and Poldrack, R. (2007) "The neural basis of loss aversion in decision-making under risk", *Science*, 315, 515–18.

Tversky, A. and Kahneman, D. (1992) "Advances in prospect theory: Cumulative representation of uncertainty", *Journal of Risk and Uncertainty,* 5, 297–323.

Wu, G. and Gonzalez, R. (1996) "Curvature of the probability weighting function", *Management Science,* 42, 1676–90.

Wu, G. and Gonzalez, R. (1999) "Nonlinear decision weights in choice under uncertainty", *Management Science,* 45, 74–85.

Chapter 13
Brain Processes During Expert Cognitive-Motor Performance: The Impact of Mental Stress and Emotion Regulation

Bradley D. Hatfield

Amy J. Haufler

Overview

The Soldier of the twentyfirst century will face unprecedented challenges regarding information management, decision-making, and adaptive motor responses on the battlefield critical to both mission success and survival. This scenario is based on the premise that the United States armed forces currently possess and will likely continue to acquire the most advanced technologies ever developed for use in the history of warfare. Such technological advancement will place a premium on the human operator's (specifically, the Soldier's) functional attention capacity in order to realize the advantages and exploit the limits of these technologies. The elicitation of intense emotional states and experience of uncontrolled arousal under battle conditions can consume, undermine, and degrade the critical mental resources and processes that future force warriors will require to execute their responsibilities. Therefore, the understanding and promotion of effective mental states, achieved through effective training programs and emotion regulation, will help the Soldier to manage arousal, preserve attentional capacity, and focus while employing such sophisticated, but attention-demanding technologies while "under fire." Such mental states may also confer resilience to the long-term consequences of stress such as post-traumatic stress disorder (PTSD). The recognition of the essential role that mental processes play in the effectiveness and wellbeing of military personnel has recently been formalized in the Army's initiative concerning the Human Dimension. This initiative represents a new development beyond the traditional human factors perspective of operator effectiveness in interaction with technology and equipment. This new element of personnel training explicitly addresses the ethical/moral, social, psychological/cognitive, and physical training elements of the Soldier's preparation as a focus complementary to the advancement of technology. This program, which was initiated in the fall of 2008 by Training and Doctorine Command (TRADOC), was described in one of the featured sessions in the second annual conference on "Sustaining Performance

Under Stress" (Adelphi, Maryland: February 16–18, 2009) with a presentation by BGEN Joseph Martz.

The discipline of Kinesiology, which centers on the scientific study of physical activity from an integrative perspective, is well suited to inform the military on the scientific basis of superior human performance in light of its established sub-specialties in biomechanics, motor behavior (such as development, learning, and control), exercise physiology, and the psychological and sociological/cultural aspects of human movement and psychomotor skill. The field of kinesiology addresses both health-related and performance-related aspects of physical activity and motor performance, with the former emphasis exemplified by studies of the role of exercise in mental health and obesity, while the latter emphasis is exemplified by studies on the execution of cognitive-motor skills under conditions of mental and physical stress. The focus of this chapter falls in the area of the cognitive-motor neuroscience perspective of kinesiology and approaches the study of superior performance guided by a basic principle that we refer to as psychomotor efficiency (Hatfield and Hillman 2001). In essence, this principle refers to the refinement and "simplification" of central neural processes with skill learning, which facilitates cognitive-motor performance by reduction of nonessential regional brain activity and attenuation of "noisy" cortico-cortical communication or networking, particularly as related to functional connectivity between the motor planning and control regions in the brain. Such a change in cortical dynamics associated with learning then translates or emerges as greater consistency and stability of motor performance (Deeny et al. 2009). That is, the reduction of neuromotor "noise" in the brain is then reflected in the quality of skeletal motor unit activity and limb action by the human operator such that the actual performance is reflective of intention. See Figure 13.1.

$$\text{Efficiency} = \frac{\text{Work}}{\text{Effort}} = \frac{\text{Psychomotor Behavior}}{\text{Neural Resource Allocation}}$$

Figure 13.1 Expression of efficiency in terms of physical work and psychomotor behavior

Such a theoretical notion has practical relevance, as the challenge to the Soldier is the maintenance or preservation of an optimal mental state under conditions of mental and/or physical stress. The research described below provides evidence that superior human performance is characterized by refined and efficient brain activity. We also provide evidence that mental stress results in heightened activation and networking in the brain, which essentially reverses the state associated with optimal performance, and, in effect, the expert's brain processes under stress more closely resemble those of the novice performer. In this stressed state, the actual performance is not reflective of intention and one could say that the performer "choked" as there is a mismatch between intent and action (for example, missing the golf putt or overshooting the basket during a free throw shot in a basketball game).

Beyond such a basic principle of brain processes and human performance, we also give consideration to individual differences in emotion regulation such that some individuals are more highly challenged in their regulation of critical brain processes under stress. This phenotypic distinction may be based on biological influence and specific genotypes. For example, variation in the serotonin transporter (5–HTT) and catechol-O-methyltransferase (COMT) alleles, results in variable disposition regarding their responses to stress. In this manner, some individuals have an advantage regarding the management of emotion and attendant "noise" in the central nervous system, typically introduced during mental stress. Emotion regulation under stress can preserve the refined and economic neurocognitive processes that are conducive to superior performance. Finally, we discuss a theoretically based model of the management of mental stress as well as the benefits of social influence resulting from membership in cohesive teams or military units. This benefit is due primarily to the presence of trust and confidence placed in one's teammates that they will "back up" one individual's efforts to execute a task by another (Zak et al. 2004, Porter et al. 2003, Barnes et al. 2008). We theorize that such group cooperation promotes attentional reserve, in addition to the maintenance of brain processes that underlie efficient cognitive-motor performance, such that the quality of the Soldier's neurocognitive state is directly influenced by his or her immediate social environment. Thus, a multi-level perspective on the underpinnings of superior human performance is articulated ranging from molecular biology, to cognitive neuroscience, to social psychology.

Our concept of superior human motor performance is based on a fundamental property of dynamic skeletal muscle activity as articulated by Sparrow (1983), which he characterized as follows: "the dynamics of coordinated muscle activity are organized on the basis of principles of minimization of energy expenditure in a process of adaptation to constraints imposed by both task and environment," (Sparrow, 1983, page 237). This statement guided our thinking about the brain processes in experts, as the recruitment of the motor units in the skeletal muscles is orchestrated centrally by the motor cortex, and we wondered whether such economy of action is also characteristic of other regions of the brain. It seems reasonable that such efficient organization in the brain would then translate into the fluid and graceful quality of movement observed in skilled individuals when they perform. In this regard, it has been well accepted for some time that the progressive stages of motor learning proceed from the cognitive stage associated with the novice to that of automaticity associated with the expert (Fitts and Posner 1967). The former stage is characterized by effortful cognitive elaboration that is likely responsible for the variability and inconsistency in performance of the beginner while the latter stage is characterized by an absence of such cognitive elaboration and a seemingly effortless orchestration of the psychomotor performance. Furthermore, phenomenological accounts of the subjective experiences reported by elite athletes across a number of sport disciplines are typically represented by such phrases as "no thinking of performance", "effortless performance", "feeling of complete control", and "involuntary experience" (Ravizza 2006). In this regard, Peterson et al. (2001) also provided neuroimaging evidence of reduced activity in the temporal lobes with automaticity of skill learning. As such, a number of levels of analysis support the view that the brain processes in the superior performer are characterized by a reduction of effort. This convergence of evidence led Hatfield and Hillman (2001) and Hatfield and Kerick (2007) to posit the psychomotor efficiency theory by which they advanced the notion that high-quality psychomotor behavior was a direct consequence of efficient neural resource allocation and is explained by the notion that a refined or simplified system is likely to result in efficient and consistent muscular performance while a "noisy" or complex system is likely to result in greater variability and inconsistency of muscle action. As a consequence of this view, and as commonly observed in real life, great performers in sport and other domains are consistently good at what they do! As such, the quality of neuromuscular coordination as described by Sparrow appears to be reflective of an efficient organization of the central brain processes.

Evidence for Psychomotor Efficiency

The scientific support for psychomotor efficiency is based on a number of investigations that have employed electroencephalographic (EEG) assessment of cerebral cortical activity during the aiming period of target shooting performance as well as basic laboratory tasks also involving visuo-motor coordination. The target shooting task is ideally suited for such investigations of brain activity as the task is attentionally demanding, requires emotional control, involves high levels of motivation, provides for ease of scoring, is of relevance to the military, and, importantly, is typically performed without movement so that the introduction of motion artifact to the psychophysiological recordings is minimized. Furthermore, the strategic placement of the EEG sensors across the scalp topography allows for functional assessment of activation in the left and right hemispheres, as well as the anterior-posterior dimension, such that cognitive and emotive inferences can be made from the EEG time series recordings. See Figure 13.2.

An illustrative example of this approach was first conducted by Hatfield, Landers, Ray, and Daniels (1982), and Hatfield, Landers, and Ray (1984) who observed a dramatic shift in temporal asymmetry of EEG alpha power during the aiming period in elite world-class sharpshooters such that a progressive decline in relative left temporal activity was evident as the time to the trigger pull approached. A comparison of the temporal activation patterns observed during the early and late stages of the target aiming period with reference tasks that differentially engaged verbal-analytic and visual-spatial processing revealed similarity with the former during the early sighting period and similarity with the latter just prior to the trigger pull. Because left temporal activation is associated with detailed and explicit monitoring of task demands while right temporal activation is associated with visuospatial processing (Springer and Deutsch 1993) the temporal EEG alpha asymmetry observed during the aiming period of target shooting was interpreted as suppression of verbal-analytic processes and explicit memory processes while the maintenance of right temporal activity (such as indicated by stability or of EEG alpha power over the aiming period) appeared consistent with the maintenance of task-specific visual-spatial processes for such a task that would logically be associated with the right hemisphere. The observed EEG patterning is also consistent with the well-known concept of automaticity of skill learning, the advanced stage of motor learning associated with expertise in which performers are not consciously aware of the details of their skilled movements as described above (Fitts and Posner 1967).

Hatfield, Landers, and Ray (1984) interpreted those findings to imply that expert marksmen progressively reduce "self-talk" and noisy and nonessential cortical activity as they approach the readiness stage to pull the trigger and progressively increase their reliance on task-relevant visuo-spatial processing as would logically be demanded by the shooting task. Such an inference was based on the comparative regional activation patterns (left *versus* right hemisphere) as well as thoughtful contrasts to tasks that evoke known cognitive processes. In this manner the contrast to the reference tasks allowed for deductive inference as to the mental processes associated with superior performance and provides evidence consistent with the model of psychomotor efficiency (for example, refinement and reduction of non-essential non-motor activity). In this manner, the

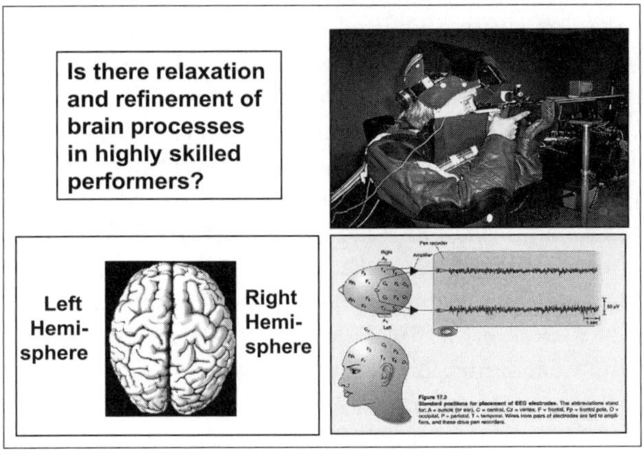

Figure 13.2 EEG can be used to assess cognitive and emotive inferences

The investigation of the question identified in the upper left quadrant can be addressed by employment of functional neuroanatomy (lower left quadrant), continuous EEG recording (upper right quadrant), and the self-paced visuomotor task as illustrated in the lower right quadrant.

spectral decomposition of EEG, with consideration of the location of the sensors in terms of the functional neuroanatomical organization of the cortex, allowed for a theoretically guided examination of the brain processes associated with superior performance. We believe that this brain state provides a useful template and biomarker or neural signature with which to characterize brain processes underlying superior performance.

Cerebral Cortical Dynamics Reflecting Brain Activation During Expert Cognitive-Motor Performance

A number of subsequent studies clearly revealed that the quality of cerebral cortical activation during precision target shooting was related robustly to the quality of performance (Bird 1987, Hatfield and Walford 1987, and Salazar et al. 1990). The participants in these early studies were comprised of elite marksmen—Olympians, members of the Army Marksmanship Unit, and National Collegiate Athletic Association (NCAA) intercollegiate competitors. Importantly, brain activity was examined under ideal and non-stressed conditions. EEG spectral power, typically in the alpha band (8–13 Hz), was contrasted between the left and right temporal regions during the aiming period of shot preparation. Because EEG alpha is inversely related to activation (Pfurtscheller 1992), mental activity was inferred from consideration of the functional neuroanatomy of the cortex and assessment of relative activity in the regions of interest from EEG power spectral analysis. Relative to the activity observed in the right temporal region (T4) of the brain, a progressive lowering of activation was typically noted in the left temporal region (T3), which was empirically indicated by higher EEG alpha power at T3 during the aiming period leading up to the trigger pull (Hatfield et al. 1982, Hatfield, Landers, and Ray 1984, and Hatfield and Walford 1987).

These early studies stimulated a number of more recent psychophysiological investigations of target shooting (Deeny et al. 2003, Haufler et al. 2000 and 2002, Hillman et al. 2000, Janelle et al. 2000a,b, Kerick, Iso-Ahola, and Hatfield 2000, Kerick, Douglass, and Hatfield 2004, Kerick et al. 2001, Loze, Collins, and Holmes 2001, and Saarela 1999) and other self-paced psychomotor performances such as dart throwing, golf, archery, and karate (Collins et al. 1990, Crews and Landers 1993, Landers et al. 1991 and 1994, Radlo et al. 2001, and Salazar 1990). Typically, these studies also involved the participation of elite athletes who had been training for their sport for many years performing under unperturbed conditions (Bird 1987, Deeny et al. 2003, Haufler et al. 2000 and 2002, Hillman et al. 2000, Janelle et al. 2000a,b, Loze, Collins, and Holmes 2001, and Salazar 1990). Such highly motivated participants offered the opportunity to observe stable cerebral cortical activity patterns that evolved over an extended period of time. Collectively, the results were consistent with the notion of psychomotor efficiency or "simplification" of neurocognitive activity with expertise. That is, precision shooting performance was achieved in skilled shooters with less cortical activation after a period of practice and motor skill learning when contrasted to novices and those of lesser shooting ability who were at the cognitive or intermediate stages of learning.

Expert—Novice Contrast of Cortical Activity

This principle was clearly supported by Haufler et al. (2000). Specifically, novice and expert marksmen were subjected to a target-shooting task as well as comparative verbal (such as word recognition) and spatial tasks (like dot localization), with which the groups were similar in terms of experience, while recording EEG. The verbal and spatial tasks were also performed in the shooting stance posture. As shown in the three panels below, lower cortical activation levels in the cerebral cortex were observed in the experts during the aiming period of shooting, as measured by gamma 40-Hz power, while no differences were revealed during the comparative tasks. The left-sided and middle panels of Figure 13.3 represent comparative log-transformed gamma power (36–44 Hz), which is also positively related to cortical activation, from the averaged homologous frontal brain regions (Panel A) and the averaged homologous temporal brain regions (Panel B). The group differences in activation associated with the frontal region also suggest that the experts were less reliant on effortful executive processing (reflecting planning and coordinating processes) as compared to their novice counterparts. Experts also revealed significantly lower beta power, a spectral band that is positively related to cortical activation, during shooting while, again, no differences were noted during the verbal and spatial tasks. Panel C shows that the experts exhibited higher levels of alpha power (8–12 Hz) at site T3 as well as lower levels of beta and gamma power. Collectively, the results clearly show task-specific relaxation in the cortex of the expert marksman.

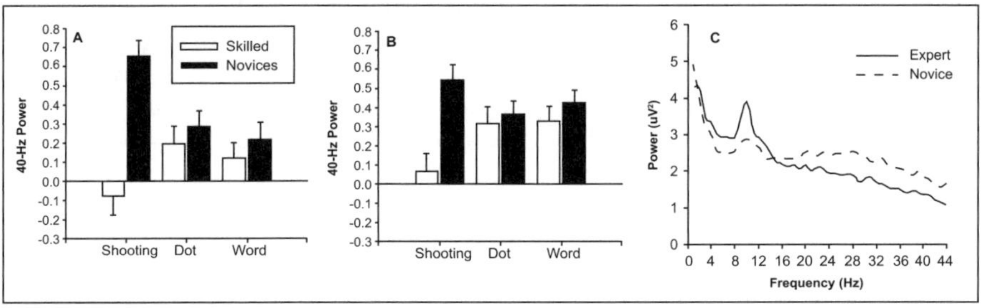

Figure 13.3 Expert and novice EEG comparisons

Comparison of expert and novice EEG gamma power for shooting and comparative tasks for the frontal region (panel A), temporal region (panel B), and spectral power (1–44 Hz) for expert and novice shooters at site T3.

Such a difference between experts and novices has also been shown to accrue from practice and learning. This was demonstrated in a laboratory setting by Gentili et al. (2008) who employed EEG recording with participants who were challenged with a target pointing task while receiving visual feedback of their performance that was distorted. As such, the study participants had to learn a new visuomotor map of the relationship between their arm action and the visual feedback from the environment. Over the course of numerous practice trials a remarkable reduction in cortical activity was observed characterized by progressive synchrony of alpha power. The relative relaxation of cortical activity over time was also associated with heightened speed, accuracy, and consistency of the pointing trajectories. In this manner, the kinematic qualities of the movement reflected the state of the brain. Such a change in cerebral cortical activity as a result of practice was also noted by Kerick, Douglass, and Hatfield (2004) who assessed EEG over the course of three months of pistol shooting training in midshipmen at the United States Naval Academy. Again, a remarkable and progressive elevation of EEG alpha power during the aiming period was observed as a function of practice over the training period that coincided with increased accuracy of target shooting performance. The regional change in activity (an observed relative relaxation over time) was most pronounced in the left temporo-parietal regions of the scalp topography.

Although such a change in cerebral cortical activity with skill acquisition could be due to several factors such as increased confidence, motivation, stress reduction, etc., evidence of the causal influence of cortical dynamics on performance has been provided by neurofeedback. As stated above, Landers et al. (1994) conducted the only study published to date in which biofeedback (left temporal slow-wave activity) was used to alter brain activity in an attempt to facilitate target-shooting performance. Based on the results of Hatfield, Landers, and Ray (1984), Hatfield and Walford (1987) and Salazar et al. (1990) 24 pre-elite archers underwent one of three treatment conditions in which one group received during a single session "correct" feedback to reduce left hemispheric (specifically, temporal) activation, a second group received "incorrect" feedback to reduce right hemispheric activity, and a third group rested and received no feedback. After one feedback session, approximately one hour in duration, comparison of pre-test and post-test performance scores revealed that only the "correct" feedback group improved target-shooting accuracy while the "incorrect" group declined in performance. The rapidity of training effectiveness is consistent with the findings of other neuroscientists who have reported very short time periods for experience-induced synaptic change in the brain (Quinlan et al. 1999).

A more recent study by Hung et al. (manuscript A, in process) provided causal evidence of the link between cortical relaxation and the quality of cognitive-motor performance. Based on the findings of the extant literature described above, Hung et al. provided targeted neurofeedback training to expert pistol shooters during the aiming period such that the study participants attempted to reproduce the level of alpha power in the left temporal region that they exhibited during their best shots or performances that they generated during a pre-test assessment period. In other words, the participants tried to mimic the regional EEG alpha power that they exhibited during their best trials. After 16 training sessions, each involving 30 one-minute trials in which auditory feedback was given upon attainment of the goal, those participants in the experimental group exhibited significantly higher alpha power during a post-test target shooting challenge relative to control group members and also achieved superior accuracy. That is, the relaxation or refinement of the cerebral cortex influenced the motor performance in a positive manner.

Cortical Coherence and Shooting Performance

Additional insight can be attained into the neurobiology of the skilled performance state by examination of functional interconnectivity or cortico-cortical communication between specified topographical regions of the brain. Such "networking" activity can be quantified by deriving coherence estimates between selected pairs of electrodes or recording sites (Busk and Galbraith 1975). In a recent study Deeny et al. (2003) assessed inter-electrode coherence between motor planning (Fz) and association areas regions of the brain during skilled marksmanship by monitoring EEG at sites F3, F4, T3, T4, P3, Pz, P4 as well as the motor cortex (C3, Cz, C4) and visual areas (O1 and O2). Coherence was assessed during a 4-second aiming period just prior to trigger pull in two groups of participants who differed in ability level. The superior performance group was labeled "experts" while the other group was labeled as "non-experts." An important dimension of the study was that both groups were highly experienced (approximately 18 years of experience in each group) but the experts consistently scored higher under the stress of competition according to group histories. Figure 13.4 illustrates the left hemisphere Fz–F3, Fz–C3, Fz–P3, Fz–T3, and Fz–O1 coherence estimates contrasted between the two groups.

A significant difference between the groups was detected for the Fz–T3 alpha band coherence, as experts revealed significantly lower values, although no other differences were observed for either the left or right hemisphere. The general lack of group differences in cortical networking seems reasonable as both were similarly experienced with the task and challenged in a similar manner. The same pattern of findings was also observed for beta band (13–22 Hz) coherence. However, the Fz–T3 results were interpreted to mean that the experts were able to limit the communication between verbal-analytic and motor control processing, thereby simplifying motor planning. The potential importance of this refined networking in the cerebral cortex is the reduction of potential interference with task-relevant cognitive and motor functions for superior shooting performance from irrelevant associative and affective (limbic) processes.

More recently, employing an extreme-groups contrast, Deeny et al. (2009) contrasted cortico-cortical communication between a number of scalp locations and the motor planning region in expert *versus* novice target shooters and observed lower coherence across the entire scalp topography in the experts. Importantly, the findings were in accord with the principle of psychomotor efficiency and, furthermore, the magnitude of the coherence estimates at a number of sites was positively associated with movement variability of the aiming trajectories. That is, higher coherence or functional communication with the motor planning region was associated with more variability or lack of steadiness while lower coherence was reflected in less variability or greater steadiness!

Collectively, the results of these studies suggest that superior performance is marked by mental economy, particularly of analytical associative processes, and that pruning of excessive cortico-cortical communication between such processes and motor regions underlies enhancement and consistency of psychomotor (shooting) performance. In essence, a cognitive neuroscience model of human performance is postulated as follows: the novice is characterized as approaching a task and performing in a verbal analytical, attentionally demanding, and effortful mode with overall increased cortical arousal and relative left hemispheric activation as measured by spontaneous EEG while the expert or highly practiced individual approaches the task and performs with automaticity and efficient resource allocation, typically in a visuospatial mode, with overall decreased cortical arousal and relative right hemispheric activation as measured by spontaneous EEG.

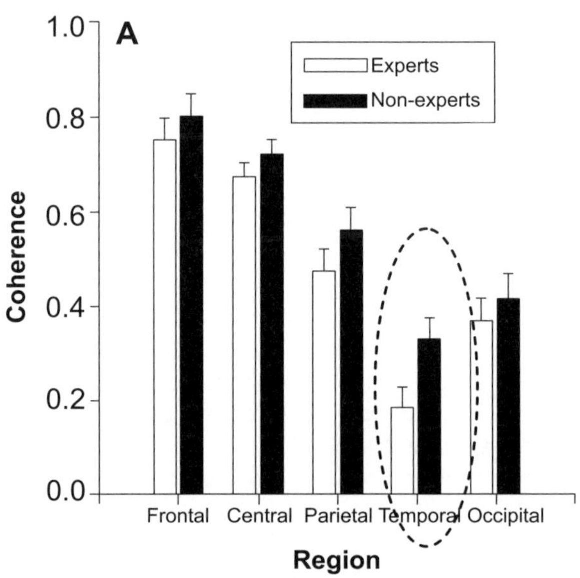

Figure 13.4 Left hemispheric comparisons of cortico-cortico communication between brain regions (frontal, central, parietal, temporal, occipital, from left to right) and motor planning area

Mental Stress and Cortical Dynamics

The neural signature for cognitive-motor expertise is characterized by a fundamental distinction between the refined organization of brain processes in the expert and the noisy cortical dynamics associated with the novice. The increased arousal associated with the uncertainty of events imposed by mental stress and the possible concern for physical harm raises the prediction that the brain state observed in experts could revert or move towards that of the novice during the experience of psychological stress. See Figure 13.5.

In a direct test of this prediction, Hung et al. (in preparation B) examined cortico-cortical communication between the temporal regions (T3 and T4) and the motor planning region (Fz) during the aiming period of a dart-throwing task under varying conditions of mental stress. Accordingly, EEG coherence for the electrode pairs (T3–Fz and T4–Fz) was assessed in 21 study participants, after completion of 36 practice sessions over a three-month period of time, during both a relatively stress-free condition in which they performed the task in isolation with no evaluation of their performance and in a relatively stressful condition in which social evaluation and comparison of their performance to others was present. The participants reported a significant increase in state anxiety during the competitive condition relative to the non-competitive condition and also exhibited a decline in accuracy of the throws. In accord with the prediction, the participants also exhibited higher functional connectivity, as indicated by higher EEG coherence, under the condition of stress, between left temporal and midline frontal regions while no such differentiation in cortico-cortical communication between conditions was evident for the right temporal region. See Figure 13.6. The investigators interpreted the findings as an increase in non-essential communication and noisy input from the left hemispheric region during mental stress that resulted in increased complexity

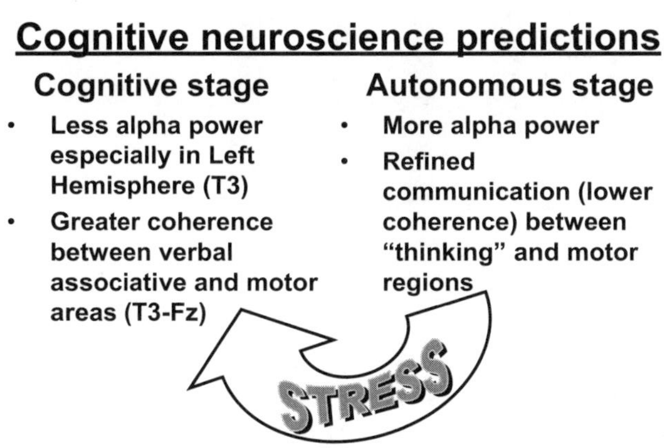

Figure 13.5 Psychophysiological correlates of the autonomous and cognitive stages of skilled motor performance

of movement planning and execution processes. Such an alteration then emerged as increased complexity, which resulted in higher variability and less consistency in the execution of the dart-throwing task.

In order to examine this possibility further (Hatfield and Haufler 2009) studied Reserve Officer Training Corps (ROTC) candidates who were exposed to mental stress in order to examine how stress impacts psychomotor efficiency by which the brain processes in the cerebral cortex are perturbed. The participants were challenged with a pistol shooting task in light of the fine motor control required, and likely susceptibility to the effects of mental stress, as well as the relevance to military interest. Based on the concept of psychomotor efficiency, which posits refined networking between non-motor and motor brain regions, and reduction of non-essential processes in superior performers (who are expert marksmen), under conditions that promote concentration and focus, Hatfield and Haufler predicted that mental stress would evoke increased networking with the motor planning regions and heightened regional cortical activation and that such stress-induced neuromotor "noise" would emerge in greater variability of the aiming trajectory and poorer shooting accuracy, relative to that observed during a low-stress condition. Beyond the work of Hung et al. (in preparation B) cited above, the significance of this work is that previous studies of cortical dynamics during cognitive-motor challenge have typically been confined to non-competitive conditions, which are likely less stressful.

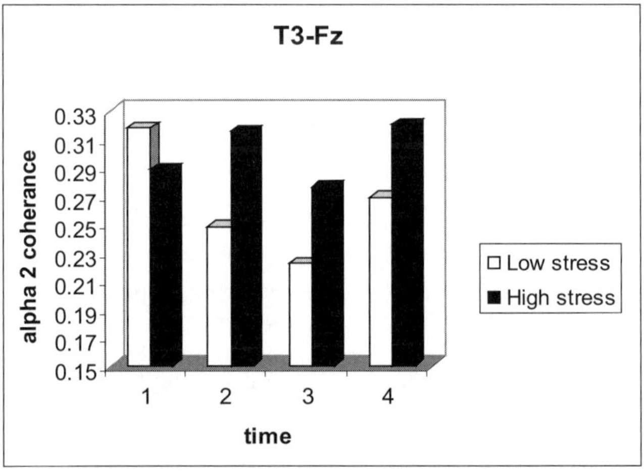

Figure 13.6 EEG Coherence for performance alone (low stress) and competition (high stress) over the four seconds leading up to the trigger pull, which was coincident with the fourth second

All participants completed both a performance-alone and a competitive condition while the order was counterbalanced. Relative to the performance-alone condition, a significant level of psychological arousal was successfully induced, albeit modest, by challenging the study participants with a competitive event during which their shooting technique and performance were evaluated by a superior officer. As such, they were subjected to social comparison, and there was public display of their performance scores to other members of the study, over the course of a 40-shot one-on-one match for accuracy (competition against an equally skilled opponent). The elicitation of elevated arousal during competition was evidenced by subjective self-report and psychophysiological assessment (heart rate, heart rate variability, skin conductance, as well as psychoendocrinological assessment such as salivary cortisol levels).

Changes in the Central Nervous System with Mental Stress

The primary measure of activation examined was the EEG spectral power in the alpha band (8–13 Hz), which was further classified into low alpha power (8–10 Hz) and high alpha power (10–13 Hz). The former is indicative of generalized arousal while the latter band is indicative of task-specific attentional processes. No differences in low-alpha power during the aiming periods prior to trigger pull were observed between the Competition and Performance-Alone conditions for low alpha, which would indicate a similar level of generalized arousal during the two conditions. However, a relative reduction or desynchrony of high-alpha power across the topography of the cerebral hemispheres was observed during the aiming period in Competition, which is indicative of enhanced attentional processing. Such a finding appears adaptive in light of the moderate elevation in arousal during Competition. Furthermore, a relative increase or synchrony was observed for high-alpha power during the aiming period (for example, approaching the trigger pull) of Performance-Alone, which would imply a progressive relaxation of regional cortical activity during aiming under the less stressful condition.

This progressive "relaxation" phenomenon was absent during the arousal induction state associated with competition. Furthermore, and consistent with our predictions, there was an increase in EEG coherence, which is indicative of increased cortico-cortical communication, during Competition. Such an elevation in coherence during the relatively aroused competitive state implies increased communication between non-motor and motor regions of the cerebral cortex and the possible introduction of neuro-motor "noise" or non-essential inputs into the motor planning system. Of note, we observed the same pattern of results for both theta and alpha band coherence. The former is thought to be influenced by hippocampal memory-related processes. In sum, and consistent with our predictions, the relative increase in theta and alpha band coherence during competition can be interpreted as more "effortful" processing under conditions of mental stress.

This interpretation was further supported by subjecting the EEG data during the respective aiming periods of Performance-Alone and Competition conditions to Independent Components Analysis (ICA), a blind separation technique that can decompose the resultant EEG record into spatially independent and temporally dynamic components or "ingredients." In essence, we observed an increase in the binding or clustering of these components during Competition with a template or base component that was specifically tied to degraded or poor performance in the aiming period during shooting. The relative clustering of components during Competition implies increased complexity of central neural processes during mental stress, in addition to the relative activation (for example, desynchrony of high alpha power) and the increased cortico-cortical communication (evidenced by elevated theta and alpha band coherence) described earlier.

As such, it appears that mental stress, induced by manipulation of the social environment and the introduction of competition, was reflected in the central and peripheral measures of arousal and indicative of a moderately aroused state. It was predicted that the increased neuromotor "noise" associated with the aroused state would translate into degraded performance as indicated by lower shooting accuracy. Inferior accuracy during Competition was not observed, but a progressive increase in movement dysfluency or "jerk" was observed during the aiming periods of this condition. Such a kinematic alteration during Competition implies effortful co-contraction of the muscles involved with arm positioning during aiming on the target. In sum, the manipulation successfully raised physiological and central arousal as expressed as heightened cortical activity and more complex brain dynamics. This altered state was expressed as more effortful performance as indicated by the kinematic analysis of the aiming trajectory.

Under such moderate arousal we observed an adaptive state—the participants were attending better as indicated by desynchrony of high alpha during competition.

In future studies, there is certainly a need to arouse participants to extreme levels of work-related stress in order to examine deleterious responses as would be associated with the battlefield and to further test theoretical notions about the impact of mental stress on cortical dynamics, decision-making, and the quality of cognitive-motor behavior. Such opportunities will be afforded with the development of new technologies for psychophysiological recording and neuroimaging that can be employed in the field. Several promising developments in imaging are occurring in this regard and one technique that offers particular promise in that of functional near-infrared spectroscopy (fNIRS), which is sensitive to hemodynamic changes in the cortex and less sensitive to motion artifact than EEG, and with which it appears quite convenient to assess activation in the frontal region of the brain in field settings. This affordance is advantageous in light of the central role that the dorsal lateral prefrontal cortex plays in emotion regulation (Northoff et al. 2004). Again, the significance of understanding the critical brain processes underlying emotion regulation lies in the possibility for preservation of efficient cortical dynamics that will enable superior performance "under pressure."

Brain Processes Underlying Emotion Regulation

Significant advancements in our understanding of the role of cortical and subcortical processes, and their linkage, in the regulation of emotion have occurred over the last decade (Davidson 2002) with the elucidation that frontal EEG recordings offer a powerful index of one's emotional state. Neurobiological and imaging work (Adolphs, et al. 1995, Davidson 2002, and Phan et al. 2002) has extended our understanding of the neural circuitry of anxiety consisting of the deep subcortical structures of the limbic system, particularly the two amygdalae structures that are essential to the orchestration of arousal-related processes throughout the brain and body, like the fight or flight response. The limbic system is a major center for emotion formation and processing, learning, and memory. The limbic system consists of the cingulate gyrus, parahippocampal gyrus, dentate gyrus, hippocampi (left and right) and amygdalae (left and right), which are represented bilaterally. The hippocampi are involved in memory storage and formation as well as complex cognitive processing, while the amygdalae are associated with forming complex emotional responses, particularly involving fear and aggression. The limbic structures are also connected with other major structures such as the cortex, hypothalamus, thalamus, and basal ganglia (Pinel 2000). Importantly, this circuitry is largely influenced by the frontal regions of the brain that can serve to inhibit or regulate fear (Northoff et al. 2004).

What remains unclear at this time are the processes involved during self-initiated emotion regulation or the control of fear. Such psychological control starts with the executive processes (inhibitory) that are largely housed in the dorsolateral prefrontal cortex (DLPFC) that, in turn, impact the activity of the emotional processes that are largely housed in the inferior regions of the frontal lobe. Such a role has been described by Ochsner et al. (2002). Importantly, the dampening of such emotion-related processes is enabled by the anatomic connection between the inferior frontal lobe and the amygdalae. This axis or pathway can be strategically controlled by the DLPFC, which would serve as a central and pivotal brain region that would exert enormous influence in the self-initiated control of fear and attendant arousal-related sequelae (Ochsner and Gross 2005, Ochsner et al. 2004). It follows that frontal and limbic activation would be inversely related during adaptive emotion regulation such that heightened frontal activity would be accompanied by diminished response in the amygdalae compared to passive fear states. An important question is if this partnership between the frontal and limbic regions is stronger in individuals who have distinguished themselves in emotionally demanding situations such as the Army Special Forces training. Beyond the dynamic activation profile described above, of heightened frontal activation and lowered limbic activation, the anatomical connections between these regions can also be captured through magnetic resonance imaging (MRI) and diffusion tensor imaging (DTI).

An important possibility is that engagement of this circuitry would afford an individual who is challenged with mental stress an adaptive emotion regulatory or management strategy by which to attenuate their emotional response and preserve the neuro-cognitive state associated with superior performance. In essence, these neural processes may underlie the ability to be "cool under pressure." Such a possibility is consistent with the theoretical position of reciprocal modulation advanced by Northoff et al. (2004). Importantly, it appears that a cognitive appraisal of situational events that are potentially threatening in terms of a task-oriented approach may provide a practical strategy with which to engage the DLPFC and down-regulate of the subcortical processes that orchestrate a fearful state. As such, a task-oriented perspective would likely involve frontally mediated executive processes and an explicit attempt to suppress or inhibit fear-related cognition.

Evidence in support of this model of emotion regulation has been gained from the classic field studies of Walter Fenz (1975) who monitored the stress responses of sport parachutists as they progressed through a series of events leading up to the exit of the aircraft at 13,000 feet. Although matched on experience, Fenz noted that the autonomic patterning of heart rate, ventilation, and skin conductance observed in superior performers was dramatically different than that of their colleagues. The former evidenced a rise in arousal that was dramatically altered and attenuated as they approached the jump, while the latter showed a monotonic elevation in arousal up to the critical jump event (while exiting the aircraft). Importantly, and consistent with the notion of reciprocal modulation in which the dorsal frontal region, which subserves executive processes, can inhibit activation in the medial frontal regions, which subserves emotional responsivity, the superior performers revealed an externally and task-oriented approach about their thought processes when interviewed while the relatively poor performers revealed a high degree of introspection and elaboration on thoughts of personal harm and injury. In this manner it appears that a task oriented cognitive approach over events that one can control may yield powerful benefits over emotional processes and facilitate the preservation of mental and physical activity that is conducive to superior performance.

In addition to such an approach, it is also theoretically plausible to employ techniques such as neurofeedback to control emotion and preserve an adaptive neurocognitive state, just as neurofeedback has been shown to enhance shooting performance, as described above. In light of the essential need to regulate the cascade of events resulting in an uncontrolled state of fear, a number of studies support the effectiveness of neurofeedback training on alterations of cortical activity in the form of specific EEG measures (Egner and Gruzelier 2004, Rosenfeld et al. 1996, Rosenfeld 1999, 2000, and Siniatchkin et al. 2000). It would seem that neurofeedback to control frontal activation (R minus L alpha power) would yield broader effects on arousal regulation and, in accompaniment with an increased number of training sessions, might yield automaticity of arousal regulation under pressure as the strategy becomes better learned. Recent studies by Allen, Harmon-Jones, and Cavender (2001) and Baehr et al. (1998) have shown that control of frontal asymmetry can be achieved in an effective manner resulting in significant alterations in emotional regulation. Importantly, the brief alterations endure for surprisingly long periods, up to five years.

Individual Differences in Stress Reactivity

The psychological state of an individual can be highly influenced by environmental factors, yet there is great variation in the abilities of each individual to cope with and subsequently perform within various environments. Individuals with strong abilities to overcome novel, stressful stimuli are often able to maintain performance levels despite frequent and dramatic changes in the environment. The ability to self-regulate or cope with environmental stimuli is governed by two factors: innate ability (such as genetics) and learned coping behaviors. While some individuals can cope adaptively to stress with relative ease, approximately 60 percent of the general population is characterized by a specific genetic variation that has recently been shown to alter the brain's ability to react to stressful conditions. Specifically, this particular genetic variant compromises serotonin transport within the brain, resulting in hyper-reactivity to emotional challenge. This heightened reactivity is a liability when the brain is working to complete complex cognitive operations required in a threatening and physically demanding environment.

The gene in question is within the serotonin pathway, which has been shown to be critical for regulation of emotional state. A gene within this pathway, the serotonin transporter gene (5–HTT), codes a protein necessary for the successful action of serotonin. Within the regulatory region of this gene exists a common genetic variation (also known as a polymorphism), of which there are two versions: the short (S) allele and the long (L) allele. All individuals carry two copies of each gene, and thus carry one of the following combinations of genetic variation: S/S, S/L, or L/L. Remarkably, carriers of either one or two copies of the short allele (that is, S/S or S/L genotypes) demonstrate brain hyper-reactivity to stressful stimuli and increased fear and anxiety-related behaviors (Hariri et al. 2002). In essence, S allele carriers may be considered "stress-prone" while LL carriers may be considered "stress-regulators." Given the extensive existing research supporting the strong association between the S allele and anxiety-related behaviors (Hariri et al. 2002, Lesch et al. 1996, Melke et al. 2001, and Ohara et al. 1998), we hypothesize that the S allele holds significant implications for mental and physical activities during performance in soldiers. The benefit to investigating the genetic influence of the stress response is in the potential for developing training interventions to better support these individuals for improved performance.

A critical point is that the individual's susceptibility to fear also holds important implications for their physical state beyond that of the brain. The psychobiology of the stress response has two main components, involving the hypothalamic-pituitary-adrenal (HPA) axis and the autonomic (sympathetic) nervous system (Sapolsky 2004). Activation of the HPA axis triggers release of glucocorticoids (including cortisol) and androgens (for example, dehydroepiandrosterone (DHEA)) from the adrenal glands into the general circulation within minutes of the stressful event. The other component involves activation of the locus ceruleus/sympathetic nervous system, triggering release of catecholamines (for example, norepinephrine) within seconds of the stressful event. Stress activates norepinephrine release in a number of regions, facilitating adaptive responses to acute stress (Koob 1999, Morilak et al. 2005). Of particular interest are findings that stress-induced elevations of norepinephrine activity promote cognitive flexibility, allowing for a switch in strategies when reward contingencies change (for example, Lapiz and Morilak 2006). This again underscores the importance of the frontal region for managing the overall stress response.

The link between norephinephrine and the tendency to abandon one's current hypothesis and adopt a new model of contingencies in the environment (Yu and Dayan 2005) is consistent with a recent hypothesis that the locus coeruleus-autonomic (sympathetic) nervous system controls the balance between optimizing performance in the current task (exploitation) and responding to other opportunities (exploration) (Aston-Jones and Cohen 2005). In addition, important psychological insights can be achieved via salivary testosterone, estrogen and alpha-amylase analysis. Leading this paradigm shift are social scientists who are testing "biosocial" alternatives to traditional models of individual differences and intra-individual change in behavior and performance (for example, Booth et al. 2003, Ennett et al. 2008, and Udry 1988). As such, it would follow that warfighters who have completed the Special Forces Assessment and Selection training program would exhibit lower overall levels of arousal with less variability (as indexed by the psychoendocrines) when challenged with the fear-eliciting stimuli as compared to those who have not completed the program.

Trust and Cooperation

In addition to individual strategies in emotion regulation, mission success typically requires cooperative effort (working in teams) as the imposed task demands, particularly during combat operations, will typically exceed the information processing and response capacity of any one individual. The management of arousal through membership in effective teams may help to preserve skilled performance "under fire" while performing alone could lead to excess arousal and degraded performance (expert performance regresses towards novice levels under stress).

Social psychologists have provided substantial behavioral evidence of a positive effect of team cohesion on self-efficacy and group performance in competitive settings. One of the central psychological constructs that determines how people are likely to respond to change and uncertainty in their environment is their confidence (or efficacy beliefs) in terms of being able to rise to the challenge and succeed. Such perceptions are particularly critical under rapidly changing conditions and performance pressure. Bandura (1997) has proposed that in team settings, "People's beliefs in their team efficacy influence the type of future they seek to achieve, how they manage their resources, the plans and strategies they construct, how much effort they put into their group endeavor, their staying power when collective efforts fail to produce quick results or encounter forcible opposition, and their vulnerability to discouragement" (p. 478).

In this manner, the social dynamics in teams can help one to achieve superior performance relative to executing a task alone particularly under conditions of significant challenge. We postulate that one primary mechanism underlying this effect is the preservation of attention reserve. Figure 13.2 (right) illustrates the relative consumption of attention capacity during task negotiation in a performance-alone mode where there is little reserve for new challenges introduced in the environment, while attention-sparing effects are realized in the team setting leaving copious reserve available for any new "surprise" demands.

Team membership also results in purposeful neuroendocrine responses that could facilitate the mental and physical state of the Soldier. An established biomarker for such a response is that of circulating levels of the hormone oxytocin. Importantly, there are specific binding receptors in the cerebral cortex, which would likely impact cerebral cortical activation and critical neuro-cognitive processes. Zak (2008) provided recent evidence that oxytocin levels are highly related to trust, social bonding, and cooperation in organized groups. As such, this powerful hormone could act as a mediating factor between the social environment, brain processes, and behavioral performance. This benefit may be manifest as an improved ability to focus on task-relevant cues as opposed to irrelevant thoughts (of, for example, personal harm and injury). In terms of brain activation, team membership may reduce "neuro-motor noise" that affects cognitive load. Fortunately, cognitive load can be quantitatively assessed in terms of cortical activation and cortico-cortical networking, operationalized as desynchrony of EEG alpha power and increases in coherence, respectively.

Kerick, Hatfield, and Allender (2007) recently demonstrated the sensitivity of EEG recordings to incremental cognitive demands during a reactive shooting task challenge comprised of three levels of demand: (1) engaging enemy targets, (2) discrimination of enemy and friendly targets, as well as (3) a combination of target discrimination and a secondary analytical task. This study revealed suppression of alpha power during the negotiation of the enemy-friendly distinction (EF-dual task load), likely due to the requisite executive response inhibition, relative to the enemy-only single task load condition. In addition, the amplitudes and latencies of key components of event-related responses to auditory probe stimuli presented during the negotiation of a primary task challenge, can be employed to assess the consumption of attentional capacity. As such, the quantification of cognitive load can be operationalized by specific features in the EEG so that the influence of team membership on brain dynamics and cognitive load can be empirically tested. The reduction of cognitive load when working in a cooperative team environment may yield a remarkable benefit in terms of the preservation of attention reserve, thus affording the ability to react to sudden changes and surprises in the environment. See Figure 13.7.

Summary

The critical brain processes described above are essentially related to the quality of human performance. Specifically, it appears that there is support for the refinement and attenuation of activity in the cerebral cortex with learning and the acquisition of expertise. Such a development has abundant support in the literature and logically relates to facilitation of the quality of motor performance as well as the preservation of attentional resources. Although some recent reports have appeared on the impact of mental stress on cerebral cortical dynamics as related to cognitive-motor performance, this area of research is undeveloped. One of the major challenges to study in this area is the elicitation of significant stress as experienced by the Soldier in the field. The conservation of attentional resources and the reduction of nonessential contributors to cognitive load would be critical to the execution of adaptive responses and effective performance. This would hold particular importance when sudden and remarkable increases in cognitive demand occur during "surprise" or emergency conditions. On the other hand, the failure to maintain attentional reserve under duress would result in an overload of attentional capacity and the failure to process critical cues.

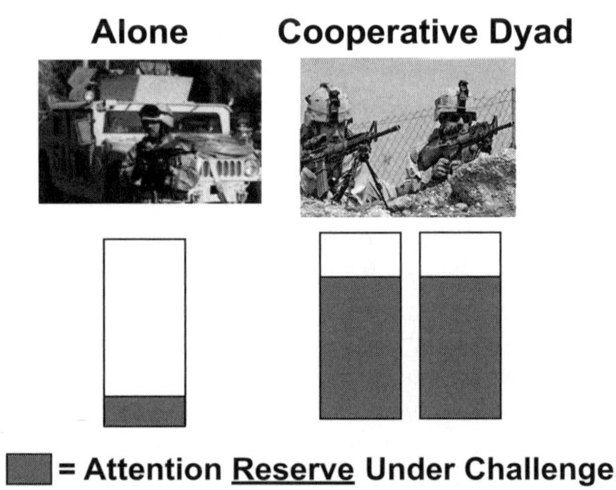

Figure 13.7 Illustration of the comparative magnitude of attention reserve under performing alone and with a cooperative dyad

However, the opportunity to observe the impact of extreme stress on brain processes will largely be enabled by the advent of new technologies that can be employed in the field. Of course, such technical developments must be married to useful theoretical perspectives that can guide data collection and allow for meaningful insights into the critical brain processes that are affected by stress and the manner with which to control such perturbations.

A number of these possibilities were developed and described above, as well as the call to consider individual differences, as related to genetic and personality factors, in their response to mental stress. Yet, despite the fact that the American Warfighter must often make important decisions while experiencing stress, there has been little research on how individual differences in the stress response influence decision processes. Previous research indicates that stress is an important moderator of neural processes involved in learning about and evaluating risk and reward. Importantly, it may be that the employment of theoretically based cognitive and neurofeedback-based strategies designed to enhance performance and regulate arousal may promote remarkable effects in certain individuals who are stress-prone. In addition, further understanding of the manner in which cohesive teams impact upon brain processes will extend the capability of enhancing human performance. The remediation of the negative consequences of mental stress on brain processes and the quality of performance may yield great benefit to the military as many of these individuals may hold invaluable skills in terms of cognitive decision-making, physical stamina and endurance, specialized areas of expert knowledge, etc. In this manner the study of expertise, individual differences, and emotion regulation may hold practical significance for the military in terms of helping highly motivated individuals to achieve excellence and overcome obstacles to their advancement.

References

Adolphs, R., Tranel, D., Damasio, H. and Damasio, A.R. (1995) "Fear and the Human Amygdala", *J Neurosci*, 15, 5879–91.

Allen, J.J., Harmon-Jones, E. and Cavender, J.H. (2001) "Manipulation of Frontal EEG Asymmetry through Biofeedback Alters Self-Reported Emotional Responses and Facial EMG", *Psychophysiology*, 38, 685–93.

Aston-Jones, G. and Cohen, J.D. (2005) "An Integrative Theory of Locus Coeruleus-Norepinephrine Function: Adaptive Gain and Optimal Performance", *Annu Rev Neurosci*, 28, 403–50.

Baehr, E., Rosenfeld, J.P., Baehr, R. and Earnest, C. (1998) "Comparison of Two EEG Asymmetry Indices in Depressed Patients v Normal Controls", *Int J Psychophysiol*, 31, 89–92.

Bandura, A. (1997) *Self-Efficacy: The Exercise of Control*, (New York City, NY: Freeman).

Barnes, C.M., Hollenbeck, J.R., Wagner, D.T., DeRue, D.S., Nahrgang, J.D. and Schwind, K.M. (2008) "Harmful Help: The Costs of Backing-up Behavior in Teams", *J Appl Psychol*, 93, 529–39.

Bird, E. (1987) "Psychophysiological Processes During Rifle Shooting", *International Journal of Sport Psychology*, 18, 9–18.

Booth, A., Johnson, D.R., Granger, D.A., Crouter, A.C. and McHale, S. (2003) "Testosterone and Child and Adolescent Adjustment: The Moderating Role of Parent-Child Relationships", *Dev Psychol*, 39, 85–98.

Busk, J. and Galbraith, G.C. (1975) "EEG Correlates of Visual-Motor Practice in Man", *Electroencephalogr Clin Neurophysiol*, 38, 415–22.

Collins, D., Powell, G. and Davies, I. (1990) "An Electroencephalographic Study of Hemispheric Processing Patterns During Karate Performance", *Journal of Exercise and Sport Psychology*, 12, 223–43.

Crews, D.J. and Landers, D.M. (1993) "Electroencephalographic Measures of Attentional Patterns Prior to the Golf Putt", *Med Sci Sports Exerc*, 25, 116–26.

Davidson, R.J. (2002) "Anxiety and Affective Style: Role of Prefrontal Cortex and Amygdala", *Biol Psychiatry*, 51, 68–80.

Deeny, S., Hillman, C., Janelle, C. and Hatfield, B. (2003) "Cortico-Cortical Communication and Superior Performance in Skilled Marksmen: An EEG Coherence Analysis", *Journal of Exercise and Sport Psychology*, 25, 188–204.

Deeny, S.P., Haufler, A.J., Saffer, M. and Hatfield, B.D. (2009) "Electroencephalographic Coherence During Visuomotor Performance: A Comparison of Cortico-Cortical Communication in Experts and Novices", *J Mot Behav*, 41, 106–16.

Egner, T. and Gruzelier, J.H. (2004) "EEG Biofeedback of Low Beta Band Components: Frequency-Specific Effects on Variables of Attention and Event-Related Brain Potentials", *Clin Neurophysiol*, 115, 131–9.

Ennett, S.T., Foshee, V.A., Bauman, K.E., Hussong, A., Cai, L., Reyes, H.L., Faris, R., Hipp, J. and Durant, R. (2008) "The Social Ecology of Adolescent Alcohol Misuse", *Child Dev*, 79, 1777–91.

Fenz, W. (1975) "Coping Mechanisms and Performance under Stress" in *Psychology of Sport and Motor Behavior II*, Landers, D., Harris, D. and Christina, R. (eds), HPER Series 10, Penn State), 3–24.

Fitts, P. and Posner, M. (1967) *Human Performance*, (eds), (Belmont, CA: Brooks/Cole).

Gentili, R., Bradberry, T., Hatfield, B. and Contreras-Vidal, J. (2008) "A New Generation of Non-Invasive Biomarkers of Cognitive-Motor States with Application to Smart Brain-Computer Interfaces", 16th European Signal Processing Conference (EUSIPCO).

Hariri, A.R., Mattay, V.S., Tessitore, A., Kolachana, B., Fera, F., Goldman, D., Egan, M.F. and Weinberger, D.R. (2002) "Serotonin Transporter Genetic Variation and the Response of the Human Amygdala", *Science*, 297, 400–403.

Hatfield, B. and Haufler, A. (2009) "Individual Differences in Cerebral Cortical Activity During Stress: Understanding and Invervention to Enhance Shooting Performance." Technical Report, Army Research Office.

Hatfield, B. and Hilmman, C. (2001) "The Psychophysiology of Sport: A Mechanistic Understanding of the Psychology of Superior Performance" in *Handbook of Sport Psychology, 2nd Ed.*, Singer, R., Hausenblas, C. and Janelle, C. (eds), (New York City, NY: John Wiley & Sons, Inc.), 362–86.

Hatfield, B. and Kerick, S. (2007) "Brain Mechanisms Underlying Superior Sport Performance" in *Handbook of Sport Psychology, 3d Ed.*, Tenenbaum, G. and Eklund, R. (eds), J.C. Wiley Publishers), 84–109.

Hatfield, B., Landers, D. and Ray, W. (1984) "Cognitive Processes During Self-Paced Motor Performance: An Electroencephalographic Profile of Skilled Marksmen", *Journal of Sport Psychology*, 6, 42–59.

Hatfield, B., Landers, D., Ray, W. and Daniels, F. (1982) "An Electroencephalographic Study of Elite Rifle Shooters", *The American Marksman*, 7, 6–8.

Hatfield, B. and Walford, G. (1987) "Understanding Anxiety: Implications for Sport Performance", *National Strength and Conditioning Association Journal*, 9, 58–65.

Haufler, A., Spalding, T., Santa Maria, D. and Hatfield, B. (2000) "Neuro-Cognitive Activity During a Self-Paced Visuospatial Task: Comparative EEG Profiles in Marksmen and Novice Shooters", *Biological Psychology*, 53, 131–60.

Haufler, A.J., Spalding, T.W., Santa Maria, D.L. and Hatfield, B.D. (2002) "Erratum to Neuro-Cognitive Activity During a Self-Paced Visuospatial Task: Comparative EEG Profiles in Marksmen and Novice Shooters", *Biol Psychol*, 59, 87–8.

Hillman, C.H., Apparies, R.J., Janelle, C.M. and Hatfield, B.D. (2000) "An Electrocortical Comparison of Executed and Rejected Shots in Skilled Marksmen", *Biol Psychol*, 52, 71–83.

Hung, T., Woo, M., Lo, L., Haufler, A.J. and Hatfield, B. (in preparation, A) "The Influence of Anxiety on Brain Processes and Visuo-Motor Performance: The Mediating Role of Cortico-Cortical Communication", *Journal of Sport and Exercise Psychology*.

Hung, T., Woo, M., Lo, L., Haufler, A. and Hatfield, B. (in preparation, B) "The Influence of EEG Neurofeedback on Brain Processes and Quality of Visuo-Motor Performance", *Journal of Sport and Exercise Psychology*,

Janelle, C., Hillman, C.H., Apparies, R.J. and Hatfield, B.D. (2000a) "Concurrent Measurement of Electroencephalographic and Ocular Indices of Attention During Rifle Shooting: An Exploratory Case Study", *International Journal of Sports Vision*, 6, 21–9.

Janelle, C., Hillman, C., Apparies, R., Murray, N., Meili, L., Fallon, E. and Hatfield, B. (2000b) "Expertise Differences in Cortical Activation and Gaze Behavior During Rifle Shooting", *Journal of Sport and Exercise Psychology*, 22, 167–82.

Kerick, S., Iso-Ahola, S. and Hatfield, B. (2000) "Psychological Momentum in Target Shooting: Cortical, Cognitive-Affective and Behavioral Responses", *Journal of Sport and Exercise Psychology*, 22, 1–20.

Kerick, S.E., Douglass, L.W. and Hatfield, B.D. (2004) "Cerebral Cortical Adaptations Associated with Visuomotor Practice", *Med Sci Sports Exerc*, 36, 118–29.

Kerick, S.E., Hatfield, B.D. and Allender, L.E. (2007) "Event-Related Cortical Dynamics of Soldiers During Shooting as a Function of Varied Task Demand", *Aviat Space Environ Med*, 78, B153–64.

Kerick, S.E., McDowell, K., Hung, T.M., Santa Maria, D.L., Spalding, T.W. and Hatfield, B.D. (2001) "The Role of the Left Temporal Region under the Cognitive Motor Demands of Shooting in Skilled Marksmen", *Biol Psychol*, 58, 263–77.

Koob, G.F. (1999) "Corticotropin-Releasing Factor, Norepinephrine, and Stress", *Biol Psychiatry*, 46, 1167–80.

Landers, D.M., Han, M., Salazar, W., Petruzzello, S.J., Kubitz, K.A. and Gannon, T.L. (1994) "Effects of Learning on Electroencephalographic and Electrocardiographic Patterns in Novice Archers", *International Journal of Sport Psychology*, 25, 313–30.

Landers, D.M., Petruzzello, S.J., Salazar, W., Crews, D.J., Kubitz, K.A., Gannon, T.L. and Han, M. (1991) "The Influence of Electrocortical Biofeedback on Performance in Pre-Elite Archers", *Med Sci Sports Exerc*, 23, 123–9.

Lapiz, M.D. and Morilak, D.A. (2006) "Noradrenergic Modulation of Cognitive Function in Rat Medial Prefrontal Cortex as Measured by Attentional Set Shifting Capability", *Neuroscience*, 137, 1039–49.

Lesch, K.P., Bengel, D., Heils, A., Sabol, S.Z., Greenberg, B.D., Petri, S., Benjamin, J., Muller, C.R., Hamer, D.H. and Murphy, D.L. (1996) "Association of Anxiety-Related Traits with a Polymorphism in the Serotonin Transporter Gene Regulatory Region", *Science*, 274, 1527–31.

Loze, G.M., Collins, D. and Holmes, P.S. (2001) "Pre-Shot EEG Alpha-Power Reactivity During Expert Air-Pistol Shooting: A Comparison of Best and Worst Shots", *J Sports Sci*, 19, 727–33.

Melke, J., Landen, M., Baghei, F., Rosmond, R., Holm, G., Bjorntorp, P., Westberg, L., Hellstrand, M. and Eriksson, E. (2001) "Serotonin Transporter Gene Polymorphisms Are Associated with Anxiety-Related Personality Traits in Women", *Am J Med Genet*, 105, 458–63.

Morilak, D.A., Barrera, G., Echevarria, D.J., Garcia, A.S., Hernandez, A., Ma, S. and Petre, C.O. (2005) "Role of Brain Norepinephrine in the Behavioral Response to Stress", *Prog Neuropsychopharmacol Biol Psychiatry*, 29, 1214–24.

Northoff, G., Heinzel, A., Bermpohl, F., Niese, R., Pfennig, A., Pascual-Leone, A. and Schlaug, G. (2004) "Reciprocal Modulation and Attenuation in the Prefrontal Cortex: An fMRI Study on Emotional-Cognitive Interaction", *Hum Brain Mapp*, 21, 202–12.

Ochsner, K.N., Bunge, S.A., Gross, J.J. and Gabrieli, J.D. (2002) "Rethinking Feelings: An fMRI Study of the Cognitive Regulation of Emotion", *J Cogn Neurosci*, 14, 1215–29.

Ochsner, K.N. and Gross, J.J. (2005) "The Cognitive Control of Emotion", *Trends Cogn Sci*, 9, 242–9.

Ochsner, K.N., Ray, R.D., Cooper, J.C., Robertson, E.R., Chopra, S., Gabrieli, J.D. and Gross, J.J. (2004) "For Better or for Worse: Neural Systems Supporting the Cognitive Down- and Up-Regulation of Negative Emotion", *Neuroimage*, 23, 483–99.

Ohara, K., Nagai, M., Tsukamoto, T., Tani, K., Suzuki, Y. and Ohara, K. (1998) "Functional Polymorphism in the Serotonin Transporter Promoter at the SLC6A4 Locus and Mood Disorders", *Biol Psychiatry*, 44, 550–54.

Petersson, K.M., Sandblom, J., Gisselgard, J. and Ingvar, M. (2001) "Learning Related Modulation of Functional Retrieval Networks in Man", *Scand J Psychol*, 42, 197–216.

Pfurtscheller, G. (1992) "Event-Related Synchronization (ERS): An Electrophysiological Correlate of Cortical Areas at Rest", *Electroencephalogr Clin Neurophysiol*, 83, 62–9.

Phan, K.L., Wager, T., Taylor, S.F. and Liberzon, I. (2002) "Functional Neuroanatomy of Emotion: A Meta-Analysis of Emotion Activation Studies in PET and fMRI", *Neuroimage*, 16, 331–48.

Pinel, J. (2000) *Biopsychology, 4th Ed.*, (Boston, MA: Allyn and Bacon).

Porter, C.O., Hollenbeck, J.R., Ilgen, D.R., Ellis, A.P., West, B.J. and Moon, H. (2003) "Backing up Behaviors in Teams: The Role of Personality and Legitimacy of Need", *J Appl Psychol*, 88, 391–403.

Quinlan, E.M., Philpot, B.D., Huganir, R.L. and Bear, M.F. (1999) "Rapid, Experience-Dependent Expression of Synaptic NMDA Receptors in Visual Cortex *in Vivo*", *Nat Neurosci, 2,* 352–7.

Radlo, S.J., Janelle, C.M., Barba, D.A. and Frehlich, S.G. (2001) "Perceptual Decision Making for Baseball Pitch Recognition: Using P300 Latency and Amplitude to Index Attentional Processing", *Res Q Exerc Sport*, 72, 22–31.

Ravizza, K. (2006) "Increasing Awareness for Sport Performance" in *Applied Sport Psychology: Personal Growth to Peak Performance*, Williams, J. (ed.), (Boston, MA: McGraw-Hill), 228–39.

Rosenfeld, J.P. (1999) "Applied Psychophysiology: Exclusions and Inclusions", *Appl Psychophysiol Biofeedback*, 24, 33–34; Discussion 43–54.

Rosenfeld, J.P., Baehr, E., Baehr, R., Gotlib, I.H. and Ranganath, C. (1996) "Preliminary Evidence that Daily Changes in Frontal Alpha Asymmetry Correlate with Changes in Affect in Therapy Sessions", *Int J Psychophysiol*, 23, 137–41.

Rosenfeld, J.P. (2000) "An EEG Biofeedback Protocol for Affective Disorders", *Clin Electroencephalogr*, 31, 7–12.

Saarela, P. (1999) "The Effect of Mental Stress on Cerebral Hemispheric Asymmetry and Psychomotor Performance in Skilled Marksmen", Unpublished Dissertation, (College Park, MD: University of Maryland).

Salazar, W., Landers, D.M., Petruzzello, S.J., Han, M., Crews, D.J. and Kubitz, K.A. (1990) "Hemispheric Asymmetry, Cardiac Response, and Performance in Elite Archers", *Res Q Exerc Sport*, 61, 351–9.

Sapolsky, R. (2004) "Stress and Cognition" in *The Cognitive Neurosciences*, Gazzaniga, M. (ed.), (Cambridge, MA: The MIT Press), 1031–42.

Siniatchkin, M., Kropp, P. and Gerber, W.D. (2000) "Neurofeedback—the Significance of Reinforcement and the Search for an Appropriate Strategy for the Success of Self Regulation", *Appl Psychophysiol Biofeedback*, 25, 167–75.

Sparrow, W.A. (1983) "The Efficiency of Skilled Performance", *J Mot Behav*, 15, 237–61.

Springer, S. and Deutsch, G. (1993) *Left Brain, Right Brain (4th Ed.)*, (New York City, NY: W.E. Freeman and Company).

Udry, R. (1988) "Biological Predispositions and Social Control in Adolescent Sexual Behavior", *American Sociological Review*, 53, 709–22.

Yu, A.J. and Dayan, P. (2005) "Uncertainty, Neuromodulation, and Attention", *Neuron*, 46, 681–92.

Zak, P.J. (2008) "The Neurobiology of Trust", *Sci Am*, 298, 88–92, 95.

Zak, P.J., Kurzban, R. and Matzner, W.T. (2004) "The Neurobiology of Trust", *Ann NY Acad Sci*, 1032, 224–7.

PART 4
Guidance from Military Leadership and Ethical Considerations

Chapter 14
Military Operations: Humans Not Machines Make the Difference

James L. Merlo

Introduction

> "Sending Americans into battle is the most profound decision a president can make. The technologies of war have changed. The risk and suffering have not."
>
> Bush, G.W. (2003)

When diplomacy fails, the United States has for over two hundred years been able to call its young men and women to arms and place them in harm's way to carry out the nation's government policies. While these goals and objectives of the government can be accomplished in any number of ways, combat often represents the final option, regardless of time zone, season, or weather. The days of well-defined lines of demarcation between the strategic, operational and tactical levels of war are all but gone. Bad decisions made by a soldier at a tactical checkpoint have the potential to be instantly aired by the international press and media and to be evaluated into an international incident in real time or at a minimum overnight, which then affects the strategy of the whole engagement. The strategic environment constantly changes, most recently from major super powers staring each other down across a European front, to the removal of evil dictators and the supervision of the reconstruction of a government that follows in a land where western ideology is reviled as evil in its own right.

What has not changed at the tactical level is the need for small units to prosper. It is the combined strength of buddy teams and fire teams that provide the foundation for squads (seven to ten troops) that are the building blocks for platoons, the lowest level of military groups that has a commissioned officer in charge. These small units are the building blocks of all successful larger units and arguably are the reason for success and potentially failure in any major conflict.

The strategic environment has gone from largely tanks in rolling plains to wheeled vehicles, able to move in narrow, crowded urban alleyways and busy city streets. The foot soldier on patrol has the ability to win the hearts and minds of the indigenous population, not the 52-ton vehicle leaving the primitive city streets in ruins (Scales 2008).

The present chapter focuses on the ground warrior as the trials, tribulations, and challenges of all military personnel regardless of their race, gender or military occupational skill are expected to be prepared to perform in the extreme environments, on demand, and in conditions that are not always conducive to maximize human performance. Such occupations range from mortar man to mechanic, from infantryman to information systems technician, and from cook to clarinet player. There are over 200 military occupation specialties (MOS) in the US Army with similar numbers in each of the sister services and each are expected to perform to their best standard regardless of the environmental disturbance or hostile activity that they might find themselves in.

Our modern-day soldiers face a collection of fierce challenges. In this chapter the challenges of humans involved in military operations are presented. This discussion begins with an overview of the five essential factors inherent in military actions: sensing, presenting, processing, communicating and understanding the situation; followed by how research can improve human performance in the intense environment of military operations.

Sensing

The human senses are robust and possess great abilities to adapt to their surroundings. For example, the haptic system gives one the ability to carefully hold a noncombatant's hand and lead them to safety or fiercely grasp a combatant's body to force movement in the desired direction, all with little cognitive effort. However, soldiers are constantly being pushed to perform in ways that we were not so keenly designed. For example, covering the body with armor and heavy loads changes the human center of gravity. Soldiers sometimes carry loads in such extreme that they have trouble standing or going prone. While the Interceptor Body Armor (IBA) is absolutely essential in the protection of the individual soldier, it is heavy (21.7 lbs (9.8 kg) for a medium size) (Merlo, Szalma, and Hancock 2008). Simple arm and hand movement gestures, especially those used in nonverbal natural and military communication, become somewhat restricted.

Gloves are worn to protect the hands, limiting some of the fine manipulations that are reflexively trained in a human's everyday life. Protective goggles, hearing protection, etc. all restrict the normal physical body structures that are so meticulously built to work in harmony. For example, most of the military common night observation devices (NODs) are bi-ocular not binocular, thus not providing true three dimensional viewing as the normal retinal disparity provided by two eye viewing. This is simply economics and weight as true binocular NODs would require two light intensifying tubes increasing weight, power consumption and costs. The same thing happens when radio handsets are shoved against the ear, only one ear is left for determining auditory stimulus localization.

To supplement and augment human senses, a plethora of electronic sensors and devices are employed across the battlefield to assist the human in the sensing task. Most of this augmentation is electronic equipment that has to be constantly maintained to remove dust and survey any damage done by excess heat and other environmental factors. In addition, special equipment like vapor tracers (an explosive residue detector) and metal detectors, all particularly susceptible to the heat. Special concessions must be made to make extra batteries available, and equipment rest periods part of the mission planning. Vehicle batteries, filters, fluids all have to be constantly maintained to ensure they are performing in the arduous conditions.

Humans and technology are not the only factors affected by thermal conditions. Specially trained military canines, which no automation has ever been able to replace, are an integral part of daily missions. These dogs require more rest and water than normal when working in the extreme and the dogs' duty day often must be adjusted even more than the soldiers', as a panting dog (a dog's equivalent to perspiration) is less effective in detecting prescribed scents. The dog's natural uniform, a fur coat, is often times also covered in an outer tactical vest (OTV) specially made for canines. Additionally, a dog's pad (foot) is burned by hot black asphalt, requiring a mat or special shade requirement for where the dog must tread.

There is no way to protect a soldier from every threat on the modern battlefield. The only relatively certain way to survive the inevitable, unexpected first contact with the enemy is through sufficient passive protection like the aforementioned IBA and vehicular armor, even on wheeled platforms. The development of the 14-ton Mine Resistant Ambush Protected (MRAP) vehicle has been a tremendous life-saver from the improvised explosive devices (IEDs) but the thick bulky armor restricts vision and audition of the local surrounding. Essentially the soldier is sensing the battlefield with much if not all of his or her senses covered.

While the advances in sensing have moved at a tremendous rate in their development and proliferation in the field, the challenge does not end with sensing, the presentation of that stimulus and the quest to make data into usable information still remains a daunting task, an issue that is discussed next.

Presenting

With myriad human and machine collectors on the battlefield, the presentation of the information and data becomes as challenging as collecting. Computer screens and other displays have to have sufficient contrast and luminance adjustments to make them readable under outdoor conditions and especially in extreme light conditions, like that present in the desert heat. Environments that are subject to extreme heat also represent additional adversities. For example, dry soil and sand flying through the air, leave at a minimum, a powdery coat of dust on even the most protected equipment.

The proliferation of network-centric information systems (*a tactical internet*), wireless technologies, and lightweight portable computers has made volumes of information available to military headquarters at all echelons of command. However, recent technologies have allowed the movement of this information to actually reach the lowest levels of units, even the *individual* soldier, while he or she is in the active maneuver space (for example, Land Warrior (LW), Future Force Warrior (FFW), Maneuver Control System (MCS), Force XXI Battle Command, Brigade-and-Below (FBCB2), etc.). The technological advancement that allows the movement of data and information across such huge boundaries is impressive, however, equal time and money has not been allocated to the understanding of how to properly display such data to the user. Helmet mounted displays (HMDs), commander's digital assistants (CDAs) and paper thin wrap-able displays have all been fielded as display technologies for the individual soldier. The data and/or information that is being displayed is most often in the exact same format as it is displayed to the soldier located in the higher headquarters's tactical operations center (TOC) (a well lit room, with large display platforms being used by a soldier who has no responsible for his or her own personal security). The only concession that is routinely made for the smaller display being used by the soldier on the front line is a smaller font size (Myles 2006).

The idea of *scalable displays* has received much attention and is now being considered more thoroughly in an effort to tailor information displays to the context and environment of the user. A further definition of scalability follows.

The tailored reception and transmission of mission essential information at the appropriate level for the Soldier, to ensure mission success while maximizing the survivability and lethality through the synergistic interaction of:

- Equipment requirements;
- Appropriate cognitive workload;
- Situation awareness and understanding for oneself and others;
- Connectivity of distributed intelligent agents.

One of the integral parts of the definition is that scalability requires tailoring. One cannot assume that the information needed by the leaders in the tactical operation center is congruent with the information requirements of the dismounted Infantryman and vice versa. The equipment constraints in terms of size, weight, and power specifically dictate that the display of the dismounted warrior will be smaller and current technological constraints will produce a degradation in areas like contrast ratio, resolution, brightness, etc. Not only does the display radically change in the field, so does the means for interacting with it. For example, joysticks, keypads, a mouse, all currently require the use of the hands. Most often the hands are occupied in the transport and operation of a weapon system and are not free to handle or manipulate a visual display. Speech recognition software is an alternative that is currently being researched as a viable alternative to traditional input devices. Stress induced voice inflection, battlefield noises, limited vocabulary, and a host of other factors will have to be adequately addressed as the voice method of interactivity is pursued.

The current research in the area of scalability is attempting to develop some design guidelines to inform design of military displays as the available real estate of the display decreases and the requirements and context of the user change. While Wickens, Gordon, and Liu (1997) offer thirteen general guidelines for displays, they place the guidelines into four distinct categories: perceptual, mental models, human attention, and human memory. These categories will have to be carefully parsed and unpacked as researchers search for the guidelines that will inform display design specifically targeting scale (such as the size of the display). Guidelines only looking at perceptual factors like contrast and resolution are going to miss the larger contribution that can be made for the display engineers. The perceptual effects of display size on emergent features (Bennett, Toms, and Woods, 1993) and/or discriminability will provide much greater insight to display designers that are working with limited real estate.

Scalability will remain a well-studied topic for some time. The range of its definitions and its sources of confounds will continue to merit hours of research and design change. One way that researchers are trying to tackle the problem is adding another modality to the mix, namely the largely unused sense of the skin, touch (see Merlo, et al. 2006).

Processing

The battlefield conditions like heat, cold, sleep deprivation and a host of other stressors all have huge debilitating effects on the long-term and working memory of the warrior (Harris, Hancock and Harris 2005). All of the aforementioned challenges make the ability for the soldier to scan, focus and act extremely difficult. There are modern efforts to solve the information challenges on the modern battlefield through both technical and human factors-based innovations.

Display principles directly related to attentional factors have to consider the trade-off of information access costs, as deep menus will be required to make up for lack of display area. Deep, hidden menus require memorization for the user and often make sources that are out of sight to be also out of mind. Information that requires multiple sources of data that must be integrated will suffer if not displayed in close proximity (proximity compatibility principle, Wickens and Carwell 1995). Displaying multiple sources of information simultaneously, obviously could become problematic in a small display. Conversely, in the case of focused attention requirements, a small display could become problematic with too much being displayed, as the close proximity of information sources could cause one problems in narrowing their focus to just one source.

Information presented on a display can provide a soldier with even more utility if the display allows some sort of interaction with its response. The interaction of display size and response or input selection will inform display design beyond the investigation of each factor in isolation. Stimulus-response capability and speed accuracy tradeoff, for example, will likely all be influenced as screen size shrinks in response to either direct (touch screens) or indirect (touch pad) position controls.

Sometimes the military tasks do not require too much thought and reflection and the desire is that the task is more of a reflexive nature. Often the military close combat tactic is simplified to the three words: move, shoot and communicate, the last being the topic of the next area.

Communicating

It is often articulated cliché that 90 percent of all human interaction problems are from a lack of proper communication. This is probably even more evident in military operations. Unless amplifier assisted intercom systems are used, verbal communications on most military vehicles is virtually impossible. The mechanical means of propulsion, such as tracks on mechanized vehicles and tanks or the prop blast from a helicopter blade, make a level of noise that is damaging to the ear if not attenuated with proper hearing protection. Notwithstanding the noise while moving, simply idling most military engines and generators are a source of noise that has to be contended. Once weapons are fired, verbal communication is always limited to shouting.

Communicating in military contexts can often influence perceived wellbeing because people often make sense of their environment by assessing other people's reactions to it, including interpreting the communications and the actions of others. When noise or other circumstances limit normal oral communication, team members read and interpret the severity of a situation by focusing on the eyes of their teammates, including their leaders. This becomes problematic in the extreme sun of the desert (most soldiers wear darkened protective goggles) or at night. Thus inflection, rate of speech and volume, all become important cues in the messenger's intent. This normally leads to a situation where it is not a matter of if there will be a miscommunication; it is simply a matter of when.

The contemporary operating environment has the military performing multinational operations resulting in planning and execution, crossing several different languages and even different dialects within the same language. Fusion cells are often built within command and control headquarters to capture what would normally be routine radio traffic and translate it across the different languages and communication systems. Often military radios are not even compatible with one another. This is even more pronounced across the different services, when operations involve expedited intra-service communication.

The shortage of translators is even more troublesome in austere combat environments resulting in many small units relying on mechanical means for translation of messages and intent to the indigenous population. Devices such as phrasalators, or small devices that speak prepared phrases in the language of choice, facilitate communication. This automation obviously lacks tone, inflections and other verbal cues that facilitate normal human-to-human communication, especially in the understanding of intent.

Leaders are in a position to influence group processes like communication, task performance, socialization, and cohesion. How people are led in extreme or dangerous contexts influences not only their effectiveness in accomplishing their goals, but how the environment impacts them psychologically. Teams that endure poor leadership in combat, for example, tend to experience higher rates of post-traumatic stress among team members than those with better leadership (Jones, 1995). Extreme environments place a premium on psychological and interpersonal factors like communication that combine to influence the physical and emotional wellbeing and performance of people.

Given the environmental factors inherent in military operations, how then do men and women in militaries around the world cope and learn to thrive in extreme settings? The next section addresses ways that humans are attempting to have better understanding of the situation.

Understanding the Situation

The speed at which the battlefield now moves is similar to the degree of change seen with the invention of the telegraph and the train or rail system, and how these modern marvels changed the way war was waged. The battlefield has become extremely fluid. Common relevant operating pictures (CROP) to promote shared understanding have become the standard with systems like Command Post of the Future (CPOF), Blue Force Tracker (BFT), Force XXI Battle Command Brigade and Below (FBCB2) and recently DARPA's TGR (Tactical Ground Reporting) system (pronounced "tiger"). Many of these systems provide on the move, near real time, situational awareness to the vehicle mounted platform level. The systems share information across the array of sensors, text messages and other information with other similarly equipped units across the battlefield. The BFT network provides commanders with the ability to digitally control and monitor their subordinate unit's status and position, even when voice communication ranges have been exceeded or have been masked by terrain.

TGR is a multimedia reporting system for soldiers at the patrol level, allowing users to collect and share information to improve situational awareness and to facilitate collaboration and information analysis among junior officers. A Facebook (www.Facebook.com the social networking website that is very popular with the current 20–30-year-old generation) of sorts, with geo-referenced data, allows the small unit leader to share patrol data and information like never before, giving micro-analysis capability from curb height on the roads in the area of operation to physical and psychological characteristics of key indigenous leaders that they might interact. This enabling of small unit leaders to share information is unprecedented and it has arrived at a particularly useful time as the small units are able to make a difference and maintain momentum every time there is a change of units or responsibility within a given region. The databases have become enormous, with thousands of mounted and dismounted operations occurring daily in the current operational environment. Providing intuitive ways for the information to be displayed and innovative ways to analyze the network continue to be a challenge as the use of these and similar systems proliferate the current theater of operation. While some challenges are being met, others arise and the help that the soldier needs is evident. Now, more than ever, the research is paramount to ensure that every advantage is given to the soldier to exceed.

Research Challenges

While the physical extremes, cognitive tests and psychological challenges are pushing soldiers to their limits, there is a constant momentum to find technological answers to the warrior's challenges. As technology advancements will indeed change the tactics, techniques and procedures used in the waging of wars, renowned military strategist von Clausewitz (1984) states that war is influenced primarily by human beings rather than technology or bureaucracy. It is most often the young leader or the ubiquitous Napoleon's corporal that makes decisions at the point of the strategic spear, inevitability bringing about the tipping point for one side to win, lose or declare a draw. In the US military, the challenge is not so much in the empowerment of the young leader to make decisions (the US Army has restructured to encourage that behavior), it is the quality of the decisions that the young leader makes that now has become the point of emphasis. Many military strategists emphasize that the strength needed to win the next war will not be as much kinetically-based as it will be cognitively-based (Scales 2006).

The military community believed for years that the classical decision-making literature appropriately defined the rational decision-making of military commanders. However, the last two decades have shown a movement towards studying the naturalistic decision-making theory and a large part of the focus has been on recognition-primed decision-making (Klein 1989, and Klein and MacGregor 1987). Military leaders, like other experienced leaders, most often make decisions based on their previous experiences. These experiences can be based on actual combat, combat training centers (Joint Readiness Training Center, Joint Multinational Readiness Center, and the National Training Center), home station training or vicariously experienced through the study of past leaders in military history. It is these experiences that arguably build a military commander's intuition.

The actual unpacking of intuition is a magnanimous task as different researchers define the term in relation to their topic. Gladwell (2004) referred to the idea of intuition as "blink" or that type of decision-making made within an instant. Dijksterhuis and Nordgren (2006) define intuition as a gut feeling based on past experience and believe that intuition plays a critical role in their theory of unconscious thought. Klein suggests that intuition must precede analysis, and that battlefield commanders are told they need to trust their own judgment and intuition, but they are not told how to develop their intuition so that it is trustworthy (personal communication, September 28, 2005).

Shattuck, Merlo and Graham (2001) found that more experienced military leaders, based on time in service and rank, tended to ask for less information when making decisions, than officers with less experience. Their study of military leader decision-making (a term they labeled cognitive integration) suggested that experienced leaders seemed to show an intuition that drove them to sample a small number of particular sources, actually choosing to ignore sources they deemed not worthy of their attention. Less experienced officers sampled all of the information sources available and usually as much of each that was allowed.

What researchers do seem to agree on in reference to intuition is that it allows timely decision-making and, in the case of experts, accurate and well-informed decisions. This ability to make good rapid decisions is essential in the military. When time is available, commanders will plan in explicit detail, including branches and sequels to the current operation. However, the dynamic battlefield does not allow for all of the commander's goals to be defined in advance. Bold and audacious maneuvers will often times lend emergent goals to be discovered in the pursuit of goals defined in advance (Klein and Weitzenfeld 1978). These emergent goals can only be achieved with quick intuitive decisions made by the commander, usually under extreme time pressure and physical duress.

The time for scientists, engineers, human factors and ergonomics professionals, and researchers to produce helpful interventions for our military has never been more important. With the declaration of war on terrorism, the call to arms for human factors and ergonomics professionals, in particular, has been formally declared (Hancock and Hart 2002). While the contributions that could be made by the human factors field are endless, there are particular areas that need immediate attention. The modern day soldier has to assimilate ever-greater amounts of information from an ever-widening number of sources. Combat conditions can impose significant demands on Soldiers' senses, limiting their ability to communicate through normal auditory and visual pathways. Noisy (weapon fire, vehicle engines) and murky (smoke, sandstorm) conditions can hinder the ability to communicate critical battlefield data. Heat, cold, sleep deprivation and a host of other stressors all have huge debilitating effects on the long-term and working memory of the warrior (Harris, Hancock and Harris, 2005). All of the aforementioned challenges make the ability of the soldier to scan, focus and act extremely difficult. There are modern efforts to solve the information challenges on the modern battlefield through both technical and human factors-based innovations.

The challenges of integrating information to make timely battlefield decisions do not reside at one particular echelon of the soldier chain of command. While the role and tasks of the soldier have evolved over time, the mission essential tasks of the soldier have not changed. Fighting and winning any nation's wars requires the translation of a government's will onto the populace of a foreign land.

Initiatives are underway to develop simulations and trainers that provide military leaders with the ability to undergo and accumulate life experiences primarily through the use of virtual simulations. These simulations could be designed so that they will adapt and respond to soldiers in an intelligent manner and portray cognitively, culturally, and intellectually accurate and challenging scenarios that are focused on identifying, developing, improving, and assessing intuitive decision-making skills in military leaders. The development of these types of decision-making simulators will provide military leaders with the ability to learn and train with scenarios that will provide life experiences, bloodlessly. Increasingly human factors—the cognitive, cultural and intellectual aspects of human conflict—are proving to be the vital factors in determining success on the battlefield (Scales 2006). For example, a well-designed interface that elicits personal interaction could lead to a self-referent memory approach by the trainee, potentially leading to a greater ability to have accurate recall when a similar situation is faced (Rogers, Kuiper, and Kirker 1977). This type of interaction with the simulator supports the theory of recognition-primed decision-making (Klein 1989) and if properly exploited with well designed interfaces will lead to perceptual learning in the areas of attentional weighting, stimulus imprinting, differentiation, and unitization (Goldstone 1998).

While there is a positive transfer of training expected from this virtual experience, a host of other benefits can be achieved from a well-made decision trainer. The pitfalls of certain heuristics and biases are well known, from the framing of decisions (Tversky and Kahneman 1981) to the readiness to use what is available to the memory or the availability heuristic (Kahneman, Slovic, and Tversky 1982). Instructing leaders on the dangers of these types of cognitive shortcuts or strategies that are used both consciously and unconsciously will potentially make a better decision maker, or at least a more informed one, especially under extreme conditions when physical and cognitive resources are potentially at their maximum limits.

The *in extremis* core is, however, one of the least studied elements of the human dimension of the military, and physical realities make it one of the most difficult circumstances for the application of science. Few researchers have endured the risk or inconvenience of studying human processes in the presence of danger, preferring instead less meaningful post hoc strategies. Nonetheless, there is no greater moral imperative for the military than understanding the *in extremis* core of the human dimension, because of the immediate and direct impact that the resultant knowledge will have on human lives (Kolditz 2007). People in dangerous contexts usually operate in teams where cohesion, communication, and leader influence are both important for survival, and take unique form and character.

As in other extreme occupations and activities, training is critical to teach warriors how to adapt to the demands of the battlefield. Training how to effectively deal with stress, for example, can mean the difference between success or failure in the heat of battle. In militaries around the world, men and women receive extensive training via many approaches. For instance, intense training in realistic environments that are similar to the places they fight are helpful to subsequent real-world performance. Warriors train in simulated war games in places like the Joint Multinational Readiness Center (JMRC) located in Hohenfels, Germany, the Joint Readiness Training Center (JRTC) located in Fort Polk Louisiana, or in the National Training Center (NTC) located in Fort Irwin, California. These mock battles at the Combined Training Centers are arguably one of the best preparations the US forces undertake before moving to hostile areas. The benefit of these training centers has been described by US military leaders as invaluable as they perform exercises that hone their skills, bloodlessly, before the real danger starts. Furthermore, many of the US military schools have been described as harder or more stressful than actual combat.

Summary and Conclusions

Technology continues to influence tactics on the battlefield. However, the tactics, techniques, procedures, and doctrine only change as a direct result of the coupling of humans with the technology. Wars will continue to be started and finished by humans. While philosophers predicted the advances in technologies would result in a higher number of casualties, we are actually seeing a reduction in the number of lives lost before nation states terminate sanctioned warfare. The skillful integration of human and machine results in better performance in battle, which in the end can save the lives of both combatants and non-combatants alike.

The intestinal fortitude and resilience of the military professional are remarkable indeed. From not only surviving but thriving in harsh weather and terrain, operating on minimal sleep and nutrients, to making cognitive decisions that have life and death implications, for themselves and others, the ability to self select into the current all volunteer force means that one welcomes this type of lifestyle and embraces its challenges. Politicians, contractors and even the media struggle to fully understand the trials and tribulations experienced by the military. A shrinking military force structure means that veterans are not physically represented in many political arenas as has been the case in the past and the high rate of speed at which technology and its uses are currently advancing can cause the military contractors to quickly become unfamiliar with the dynamic nature of the operations being conducted in the active theater. Dangerous and extreme environments limit media exposure and seldom can the real effects be truly understood with snapshots in time.

Without the proper application of human factors one could assume that a natural selection or Darwinist approach to proper human machine fit might evolve with time. However, armed with the knowledge of human capabilities, strengths and weaknesses that we now have, there is no reason that we have to wait for this natural progression. Instead, we must implement the human factor practice more readily and stringently into the military, preserving not only our precious warrior citizens' lives, but preserving our nation's freedom as well.

References

Bennett, K., Toms, M. and Woods, D. (1993) "Emergent Features and Graphical Elements; Designing More Effective Configural Displays", *Human Factors*, 35, 71–98.

Bush, G. (2003) "Address before a Joint Session of the Congress on the State of the Union (January 28, 2003)", *available at http://www.presidency.ucsb.edu/ws/index.php?pid=29645* .

Dijksterhuis, A. and Nordgren, L. (2006) "A Theory of Unconscious Thought", *Perspectives on Psychological Science*, 1, 95–109.

Gladwell, M. (2004) *Blink: The Power of Thinking without Thinking*, (New York City, NY: Little, Brown, and Company).

Goldstone, R.L. (1998) "Perceptual Learning", *Annu Rev Psychol*, 49, 585–612.

Hancock, P. and Hart, S. (2002) "Defeating Terrorism: What Can Human Factors/Ergonomics Offer?" *Ergonomics in Design*, 10, 6–16.

Harris, W., Hancock, P. and Harris, S. (2005) "Information Processing Changes Following Extended Stress", *Military Psychology*, 17, 115–28.

Jones, F. (1995) "Psychiatric Lessons of War" in *War Psychiatry*, Jones, F., Sparacino, L., Wilcox, V., Rothberg, J. and Stokes, J. (eds), (Falls Church, VA: US Army Office of the Surgeon General).

Kahneman, D., Slovic, P. and Tversky, A. (1982) *Judgment Under Uncertainty: Heuristics and Biases*, (New York City, NY: Cambridge University Press).

Klein, G. (1989) "Recognition-Primed Decision", *Advances in Man-Machine Systems Research*, 5, 47–92.

Klein, G. and MacGregor, D. (1987) "Knowledge Elicitation of Recognition-Primed Decision-Making", (Fairborn, OH: Klein Associates, Inc.: US Army Research Institute Field Unit, Fort Leavenworth, KS).

Klein, G. and Weitzenfeld, J. (1978) "Improvement of Skills for Solving Ill-Defined Problems", *Educational Psychologist*, 13, 31–41.

Kolditz, T. (2007) *The in Extremis Leader: Leading as if Your Life Depended on It*, (San Francisco, CA: Jossey Bass).

Merlo, J., Szalma, M. and Hancock, P. (2008) "Stress and Performance: Some Experiences from Iraq" in *Performance Under Stress*, Hancock, P. and Szalma, J. (eds), (Aldershot, England: Ashgate Publishing), 359–78.

Merlo, J., Terrence, P., Stafford, S., Gilson, R., Hancock, P., Redden, E., Krausman, A., Carstens, C.R.P. and White, T. (2006) "Communicating through the Use of Vibrotactile Displays for Dismounted and Mounted Soldiers", 25th Annual Army Science Conference, Orlando, Florida, *available at http://www.asc2006.com/posters/JP-14.pdf* .

Myles, K. (2006) "The Identification of War-Fighting Symbology with the Use of a Small Display", (Aberdeen Proving Ground, MD: US Army Research Laboratory (ARL-TR)) .

Rogers, T., Kuiper, N. and Kirker, W. (1977) "Self-Reerence and the Encoding of Personal Information", *Journal of Personality and Social Psychology*, 35, 678–88.

Scales, R. (2006) "Clausewitz and World War IV", *Armed Forces Journal, available at http://www.armedforcesjournal.com/2006/2007/1866019* .

Scales, R. (2008) "Foreword" in *Performance Under Stress*, Hancock, P. and Szalma, J. (eds), (Aldershot, England: Ashgate Publishing), xv–xvi.

Shattuck, L., Merlo, J. and Graham, J. (2001) "Cognitive Integration: Exploring Performance Differences Across Varying Types of Military Operations", Fifth Annual Federated Laboratory Symposium on Advanced Displays and Interactive Display, College Park, MD, 2001.

Tversky, A. and Kahneman, D. (1981) "The Framing of Decisions and the Psychology of Choice", *Science*, 211, 453–8.

von Clausewitz, C. (1984) *On War*, (Princeton, NJ: Princeton University Press).

Wickens, C. and Carswell, C. (1995) "The Proximity Compatibility Principle: Its Psychological Foundation and Its Relevance to Display Design", *Human Factors*, 37, 473–94.

Wickens, C., Gordon, S. and Liu, Y. (1997) *An Introduction to Human Factors Engineering*, (New York City, NY: Addison-Wesley Educational Publishers, Inc.).

Chapter 15

Is Supraphysiological Enhancement Possible, and What are the Downsides?

Karl E. Friedl

Introduction

Human Performance Optimization (HPO) involves strategies to sustain normal human performance in the face of operational stressors that degrade performance. These stressors and their countermeasures, including selection, training, feeding, rest, equipping, and leadership, have been examined and modeled in a longstanding Army research program (Friedl and Allan 2003). Biomedical Standards is a related area of active research that involves defining limits of human tolerance in order to match equipment and doctrine to human capabilities, instead of trying to change the human to fit the equipment and tactics. This is vitally important as modern technologies outstrip human operator capabilities, with intolerable biodynamic forces, virtually unlimited continuous operation without fatigue or failure, and ability to operate in environmental conditions too harsh or toxic for unprotected humans (Friedl and Allan 2003, Friedl 2008).

Human Performance Enhancement (HPE) strategies to modify human form and function to create superhuman or "supraphysiological" capabilities beyond the normal biological range, using techniques such as surgery, genetic modification, pharmacology, or neural stimulation, is not a current research focus. Most of these technological opportunities arise from research to treat disease conditions, justified on the basis of risk-benefit tradeoffs, where the risk of adverse effects is outweighed by the consequences of untreated disease. Thus, opportunities for HPE based on emerging medical technologies are generally known to those monitoring these advancements. Also known are some of the downsides to applications in healthy individuals. There are several common themes pertaining to safety issues and ethical concerns when we try to improve upon rather than optimize human physiology. This paper presents a few common examples to explore these themes.

Is More Medicine Better?

Modern medical technologies can improve quality of life and may even provide cures for injury and disease. Some of these therapies are assumed to be part of a continuum that can extend beyond restoration of normal function and provide more of the same effect with a supranormal functioning. However, biological deficiencies and excesses do not often travel on the same trajectory of biological effect. For example, growth hormone therapy to the aged provides improvements in muscle mass, reductions in body fat, and other benefits. These benefits occur in patients that have age-related reductions in circulating levels of growth hormone and suboptimal body composition, including reduced muscle mass that is insufficient to safely conduct normal everyday physical activities (sarcopenia) (Salomon et al. 1989). Some athletes have used growth hormone supplements to increase muscle mass and strength above levels in the normal human range.

In contrast to replacement therapy to correct a deficiency state, there are multiple derangements associated with adult growth hormone overproduction (acromegaly). For someone seeking performance enhancement, some of the problems are ironic, especially the weakness and abnormal muscle organization that accompanies this increase in muscle mass (Khaleeli et al. 1984). The quality of the muscle produced by excess growth hormone is different than that produced by anabolic steroid therapy, and has been described by one of the earlier pioneers in anabolic hormone effects as "soft and flabby" muscle (Kochakian and Stettner 1948). Thus, the effects of replacement therapy for someone deficient in growth hormone are very different than the effects of someone administering additional hormone to chronically boost levels above the normal biological range.

Similarly, anabolic steroids used as hormone replacement therapy for aging men, provide specific benefits to bone strength, muscle mass sustainment, and sense of well being. More hormone provided to normal healthy men further increases muscle mass and this can be accompanied by increases in strength; unlike most hormones, there is no disease of androgen overproduction that has been described for men. The abnormally huge increases that have been achieved in lean mass of body builders with a deranged self-image ("big-arexia nervosa") would not be possible without use of anabolic steroids (Choi, Pope, and Olivardia 2002). Power lifters who are focused on strength performance rather than muscle size, may actually derive their greatest advantage from increased aggressiveness in training and competition, although this effect is difficult to scientifically quantify. For military use, the behavioral changes that occur with use of anabolic steroids make this an unsuitable drug for performance enhancement, and it would not be ethical or sensible to administer to women to create masculinized women (Friedl 2005). Also not such good ideas are concepts and claims that anabolic steroids might be useful in accelerating healing of musculoskeletal injuries, "replace" deficiencies produced by high stress conditions, or act as a "brave pill" and increase motivation of soldiers (Friedl 2005). Other problems occur in accommodating abnormal muscle size on a normal human skeletal frame, such as tendon injuries and thermoregulatory challenges in hot environments (Friedl 2005).

HPE Experiments of Nature

Myostatin regulates the utilization of satellite cells to create new muscle cells and the myostatin gene presents a target for improvement on nature through genetic selection or engineering. This follows logically from the observations that a mutational deletion of the gene that codes for myostatin has intriguing consequences to human performance with tiny "Hercules" babies with great muscle mass (Schuelke et al. 2004). There is a version of cattle with mutations in the myostatin gene that are now being bred for their great mass and leanness (McPherron and Lee 1997). With such emerging discoveries, Aldous Huxley's Brave New World depiction of manipulating embryos to create a force of optimized manual laborers appears to be within reach. The amount of coal or gravel that an individual can shovel is largely dependent on the individual's mass of working muscle and on the amount of food energy they are provided (Spurr 1986); thus, a deficient myostatin gene, especially on a genetic background for a large skeletal frame might optimize a specifically cultivated herd of human laborers for a nation where people are more abundant and affordable than earth moving equipment and other technologies. The question of interest to physiologists is what deficiency or compensation occurs in the absence of myostatin. One hypothesis is that in the absence of a functioning myostatin, the unrestrained use of satellite cells may cause a depletion that leads to accelerated loss of muscle mass (and bone) later in life when cells can no longer be replaced. If this is the case, the cultivated laborers might have only a relatively short work life; these overly specialized strong mutants would be at a marked disadvantage against a normal military opponent with greater mobility, energy efficiency, and higher thermotolerance. Further investigations into the myostatin gene may lead to new treatments for age-related sarcopenia and be a great benefit to the ageing population and for individuals with muscle wasting diseases (Gonzalez-Cadavid et al. 1998). Based on recent history with performance enhancing drugs, there will be little or no lag time between the first emergence of a new promising treatment for muscle wasting and the widespread repurposing of that treatment to strength athletes and body builders.

Naturally occurring genetic mutations remind us that enhanced capabilities come with consequences. For example, brain wiring in savants that results in supernormal skills with arithmetic and memory seems to come hand-in-hand with deficient psychosocial skills (Treffert and Christensen 2006). One current explanation is that abnormal cognitive processing that involves intense "restricted and repetitive behaviors" is an alternate cognitive strategy that results from a deficiency in awareness of self not present in so-called neurotypical ("normal") individuals (Vital, Wallace and Happe 2009). The plasticity of the brain permits compensation for some extreme differences that permit survival but invariably produce other changes. In another example, sickle cell trait is protective against malaria, providing an overall advantage to populations in a malarial environment but including a subset of individuals homozygous for sickle cell trait who pay the price for the greater survival advantage of the group in the form of painful and disabling sickle cell disease (Aidoo et al. 2002). Significant shifts in physiological balance to produce special advantages appear to be accompanied by health risks and other compensatory deficits. The military should never be so anxious to be the earliest adopter of a new and incompletely tested "bright idea" that they end up providing the experimental evidence for why it was actually a bad idea.

What is the Risk-Benefit Tradeoff for Performance Enhancement?

The agricultural community has been a source of new concepts for performance enhancement drugs. For example, "nutrient partitioning agents" that are economically important in livestock production have crossed into the athletic community for performance enhancement (Prather et al. 1995, Dumestre-Toulet et al. 2002, Hopkins, Spina, and Ehsani 1996). Similar to their uses in meat product, a range of beta-2 adrenergic agonists have specific effects on muscle tissue to efficiently enhance protein accretion and reduce fat stores. One of these compounds, clenbuterol, is used as a medical treatment for human asthmatics, where it also relaxes the airways. In recent years, numerous athletes have been disqualified in international competition for clenbuterol use to enhance physical performance in sports such as swimming and cycling. The primary risks associated with these and related performance enhancing drugs are cardiac effects, and symptom complaints have occurred in populations consuming contaminated meat from animals slaughtered too close to the last drug treatments. Another adrenergic drug, ephedrine, has synergistic effects with caffeine that have been demonstrated to be especially promising for stimulation of fat metabolism. However, ephedrine (including botanical products with naturally occurring ephedra) is now banned from use in the US based on a pattern of cardiac deaths associated with use or overuse of the supplements (Haller and Benowitz 2000). In carefully regulated studies, the Canadian military continues to explore this combination for the important thermogenic benefits for soldiers operating in cold environments (Vallerand, Jacobs, and Kavanagh 1989).

More serious diseases have resulted in the development of riskier manipulation of physiological systems, such as gene therapy. Greater risks associated with treatment are outweighed by highly likely disease outcomes without treatment, such as the inevitably fatal progression of many neurodegenerative diseases. The first trials of stem cell transplantation into the brains of patients with Parkinson's Disease were successful in restoring dopaminergic secretory activity but problematic in restoring normal function for several reasons. Recent studies with more precise viral vector delivery of gene therapy have more successfully corrected specific deficits in Parkinson's patients and show great promise in disease treatment (Kaplitt et al. 2007). The science of Parkinson's Disease has been heavily supported by the US Army (Isacson and Kordower 2008, Federoff et al. 2003) which means that service members may be the early beneficiaries of validated treatments, but it also means that the Army is well aware of the state-of-the-art for medical technologies that may be repurposed by an opponent. The actual use of such medical technology to enhance a specific form of human enhancement advantageous to the military is conceivable. An example application could be to increase endogenous enzyme activity to increase inactivation of toxic chemicals, allowing troops to operate with greater freedom in contaminated areas (Saxena et al. 2006).

Do We Even Know What We Really Want?

There is a more fundamental problem in human performance enhancement in characterizing what optimizes success in military missions. And, if we cannot adequately describe sought-after performance, we are even less likely to be able to adequately measure it. The US Army targeted a seemingly simple and straightforward set of job selection standards for performance research in the 1970s and 1980s. Physical performance standards were based on the premise that enlisted military occupational specialties (MOS) could be aggregated into jobs that could be described by monotonically predominant requirements such as muscular strength (strength endurance), aerobic capacity, or low physical demands (Teves, Wright, and Vogel 1985, Haisman and Vogel 1986). After more than two decades of intensive research efforts, including field testing and various implementations of physical testing and selection policies, the primary conclusion is that specialized task selection is fraught with problems and not practical (Vogel 1999, Nevola 2009).

The basis of physical strength standards came into sharp focus when greater sex integration was mandated in the 1990s. Very clear differences in upper body strength of most women compared to most men led to questions about the validity of strength standards, especially when it was discovered that women were highly successful in all of the MOSs that were opened to women. The Canadian military forces led the world in demanding a science basis linking selection standards to job success. Their conclusion was to withdraw from all the existing physical selection standards and pursue new task-relevant testing methods that could be independent of sex bias or any other non-job related discrimination (Knapik et al. 2007). One more attempt by the US Army to link task performance abilities and job requirements focused on a practical and important performance outcome measure—physical injury (assessed by occurrences and lost duty days). Light wheeled vehicle mechanics were selected for this study based on their very high rates of musculoskeletal injury in both men and women. The study carefully assessed critical MOS tasks, measured strength requirements associated with those tasks, developed relevant tests of strength for those tasks (for example, force applied to a torque wrench), and then examined injury incidence (Knapik et al. 2007). The conclusion was that the difference between requirements and actual capabilities of soldiers performing in this specialty was not clearly related to musculoskeletal injuries.

Even when we do identify a specific performance characteristic that we could enhance, we need to make sure that it addresses a military performance need. Creatine supplements, taken in doses equivalent to the amount that would be obtained from daily consumption of impossibly large amounts of red meat, produce short burst-type strength enhancement, based on demonstrated effects to increase the energy substrate required for power generation in skeletal muscle (Casey et al. 1996, Hultman et al. 1996). The specific benefits of creatine to this particular type of physical performance enhancement are well described and understood but the benefits to a military mission have not been identified. This conclusion is supported by an Army study where strength increases produced by creatine supplementation provided no clear benefit to individuals or units involved in rigorous field training activities (Warber et al. 2000).

Attempts to simply assess mental functioning in a reliable and meaningful way have been even more difficult than physical performance standards. Neuropsychological test batteries have been investigated in the Army at least since World War I (Yerkes 1920). More than thirty years of research in automated neuropsychological testing by WRAIR has led to the 2008 adoption of an abbreviated field test battery applied to every service member before operational deployment. The objective of this testing was to establish some form of predeployment cognitive baseline against which a subsequently injured or "poorly functioning" individual could be compared as one of several tools used in diagnosis and recovery assessment from functional impairment. The larger hope for this practical application of neuropsychological testing is that return to duty standards could be established that are based both on normal ranges of variation for an individual as well as associated with military task performance (Friedl et al. 2007). Although similar tests are being applied for return to play standards for individuals involved in impact sports, there is relatively little published data on normal variations, true functional deficits, associations with serious deficits of medical concern, and least of all, associations with performance success.

As we become a more technical and sophisticated military, the actual skill sets we demand may come from the overweight nerd living in his mother's basement who knows how to hack computers. Earlier conceptions of the ideal soldier may in fact exclude the most important skill sets for nearterm military objectives. Agility of response for the military is crucial and this calls for the broadest diversity and capability sets, including possibly the overweight nerd. This doesn't necessarily mean that we should not select super capabilities but we at least need a mix of those super capabilities and cannot claim a single performer ideal. Thus, a superbly strong individual is not likely to also be the best qualified individual for endurance running; the tall strong soldier may not be a good "tunnel rat" like the shorter and smaller individual; and the superb leader with great interpersonal skills may play a very different role from the equally critical borderline autistic individual with great skills in connecting patterns to predict enemy intentions. This seems to have been recognized in the superhero comic books, where superpowers combined in a single individual ("faster than a speeding bullet, able to leap over tall buildings, more powerful than a locomotive") are now reflected in superhero teams composed of individuals with complementary superhuman skills.

The lessons in nature suggest the strategy of optimization and balance over superspecialization and supraphysiological capability. Consider the contrast between the Aye-aye and the rat. The Aye-aye is an endangered prosimian species that was once abundant in Madagascar. This species developed a very high degree of specialization, with the evolution of an oversized finger to pry insects out of holes and an egg out of the shell. In addition to an unfortunate association with a belief that they must be killed on sight to protect villages from evil spirits, these specialized primates have not been able to adapt to deforestation encroachments on their environment and food foraging technique. The rat reflects all the opposite characteristics of adapatability, hardiness, and resilience that account for its worldwide abundance. The Aye-aye may be smarter than the rat but lacks versatility. Superspecialization of soldiers could similarly restrict the agility to adapt to even more rapidly changing tactics and threats on the battlefield.

Individual Control and Response Agility

Voluntary control is an important consideration in performance enhancement, especially for adaptability to changing circumstances. Most aspects of HPO involve readily controllable or at least reversible changes, while most of the HPE strategies currently under consideration are long lived or permanent effects. A simple example of this is visual acuity. Corneal refractive surgery allows young soldiers to compete for jobs that were previously closed to individuals requiring optical correction and some individuals have sought even better than 20–20 visual acuity through this procedure (Bower 2007). It is too early to fully understand the limitations and consequences of this technology including how much of the cornea can be removed and how many times within a lifetime before the integrity of the eyeball is compromised. Contrast this HPE solution with observations in nature that suggest possibilities for voluntary control. As an example, the Moken are a group of fisherman in the Andaman Islands with extraordinary visual acuity. When Moken children dive for pearls, they perform a squinting-type of maneuver that bends the lenses at the same time that they constrict the pupils (rather than the normal response to dilate in lower light conditions underwater), optimizing to the limits of what the human eye can achieve and providing at least a two fold advantage in the size of small objects that they can detect (Gislen et al. 2003). It appears that this is a trainable phenomenon (Gislen et al. 2006). Eventually, with technology assists in the form of biofeedback equipment, it should be possible to provide "eyeball training" to change accommodation and avoid the need for spectacles. Further "eagle eye" advantages are likely with other developments in assistive equipment developed for patients with scotomas and other low vision problems but the technology has obvious adaptations to the enhancement of current capabilities in human vision. Other aspects of the technology assists such as electronic retina and gene therapy for color blindness will be likely repurposed to bionic supervisual acuity or increased receptor density or a fourth cone type to introduce infrared or ultraviolet visual capability.

Performance optimization through biofeedback training of physiological systems can provide a variety of huge advantages with limited downsides. Earlier in this paper, studies were described on the investigation of caffeine and ephedra supplements to stimulate thermogenesis in cold environments, including for purposes of reducing peripheral vasoconstriction and reduced manual dexterity (Cheung et al. 2008). An alternate HPO solution would be to train cold weather troops in a "hunting" reaction (increasing blood flow to the hands instead of vasoconstriction), just as individuals with Raynaud's syndrome are trained using biofeedback to not vasoconstrict abnormally in the face of a cold challenge (Brown et al. 1986). This is a known phenomenon in aboriginal fishermen who have been shown to vasodilate when they put their bare hands in cold water, sustaining manual dexterity to handle game and fish (Hirai, Horvath, and Weinstein 1970). Unfortunately, there are no current military research programs investigating these mental training approaches to performance enhancement, or the biomonitoring technologies that might provide a reliable tool to produce, as well as measure, training success.

A Spectrum of Choices in Fatigue Manipulation

The Army has machines that can function for many hours and days without fatigue or breakdown but is limited by human operators that must rest and if pushed beyond fairly well demarcated limits of continuous performance, hours will no longer be mentally capable even with stimulants drugs. The operators will exhibit physical exhaustion and be at increased risk of injury. Increasing duration of wakefulness leads to increasing impairment of cognitive function, starting with loss of the most complex functions such as judgment and decision-making.

Fatigue is one of several natural behavioral mechanisms that limit the organism from continued activity that may result in some kind of physiological compromise including injury and death (Friedl and Penetar 2008). It protects the musculoskeletal system from overexertion and brain and other organs from metabolic failure. Strategies to counter fatigue and delay rest fly in the face of physiological realities. Caffeine and carbohydrate supplements are provided to soldiers in supplement form because both of these usefully extend performance in sustained or continuous military operations where fatigue is a key limiter (Friedl and Hoyt 1997, Killgore, Balkin, and Wesensten 2006). A 600 mg dose of caffeine can temporarily restore performance at 72 hours of sleep deprivation to baseline performance for several hours (Rasmussen 2008). After this, the person must sleep, and, in fact, improved restorative sleep even with sleep enhancers is a current focus as only sleep provides restoration of optimal function. Amphetamines and modafinil also may be indicated to extend performance under emergency conditions but come with the risk of further impairing judgment with extending wakefulness and, in the specific case of amphetamines, possibly increasing risk. Amphetamines were well studied and tested in troops in World War II to increase endurance, and promptly abandoned by most Armies because of misuse (except the US Army). They worked but were most helpful with temporary use in emergency conditions, an approach that has been further researched and now forms officially sanctioned policy within the DoD (Tyler 1947, Caldwell 2008).

Just as night vision technologies allowed the Army to "own the night" starting more than a decade ago, anyone who could substantially moderate the fatigue problem would "own chronobiology." A previous Defense Advance Research Projects Agency (DARPA) program known as "Continuous Assisted Performance" considered novel blue sky approaches to extending performance through studies such as unihemispheric sleep in dolphins and continuous flight performance without food or sleep of migrating birds; most of these are not directly applicable to humans unless we surgically alter the brain, such as sectioning the corpus collosum and significantly alter other aspects of human function. One product from this program was the advance of research on ampakines, a class of compounds that increase cognition and may also be useful in neurodegenerative diseases (Porrino et al. 2005). Another program at the Walter Reed Army Institute of Research (WRAIR) involves genomics studies of individuals who are especially fatigue-resistant compared to those who are fatigue-susceptible. Other efforts such as the use of transcranial magnetic stimulation (TMS) to activate the frontal lobes in healthy young soldiers were also explored (Denslow et al. 2005). Changing healthy normal brains in this way with electrodes or even with "noninvasive" TMS raises the question of voluntary control, reversibility, and if there is an ethical line which the military should not cross in the spectrum from mental training and generally safe and accepted stimulants all the way through to brain implants to create fatigue unlimited soldiers.

Conclusions

The primary conclusion of this paper is that we may wish that we could improve on nature; we most certainly will influence our own evolution; but before we can enhance human performance effectively, most of what we still have to understand about human physiology is yet to be learned. The human system is currently optimized to survival in a particular environmental niche. Hochachka and colleagues (1998) have described the specialization to hypoxia that gives us capabilities that target endurance exercise capabilities, and our biochemistry, muscle physiology, etc. is suited to related specializations that stem from this capacity. The extent of our specialization is highlighted by comparison to other mammalian relatives occupying other specialized niches such as pinnipeds adapted to cold water and deep dive breath holding (Hochachka 2000) or camelids adapted to heat and severe dehydration (Schmidt-Nielsen 1967). It is imaginable that Mankind will influence its own evolution to a Heinlein-conceived science fiction creature composed primarily of brain floating in an artificial supportive environment, perhaps with unlimited lifespan. This will be a very different species from the current physical creature where mind and body are interdependent for healthy function.

References

Aidoo, M., Terlouw, D.J., Kolczak, M.S., McElroy, P.D., ter Kuile, F.O., Kariuki, S., Nahlen, B.L., Lal, A.A. and Udhayakumar, V. (2002) "Protective Effects of the Sickle Cell Gene against Malaria Morbidity and Mortality", *Lancet*, 359, 1311–12.

Bower, K.S. (2007) "Evaluation of the Safety and Efficacy of Excimer Laser Keratorefractive Surgery in the U.S. Army Soldiers Using the Latest Battlefield Technologies. Technical Report", (Washington, D.C.: Walter Reed Army Medical Center) ADA466560, 107.

Brown, F.E., Jobe, J.B., Hamlet, M. and Rubright, A. (1986) "Induced Vasodilation in the Treatment of Posttraumatic Digital Cold Intolerance", *J Hand Surg [Am]*, 11, 382–7.

Caldwell, J.A. (2008) "Fatigue Management for Military Aviation—Select Studies and Experiences in US Air Force", Technical Report, (Neuilly-sur-Seine Cedex, France: Research and Technological Organization, North Atlantic Treaty Organization) RTO-HFM/WS-151, 1–20.

Casey, A., Constantin-Teodosiu, D., Howell, S., Hultman, E. and Greenhaff, P.L. (1996) "Creatine Ingestion Favorably Affects Performance and Muscle Metabolism During Maximal Exercise in Humans", *Am J Physiol*, 271, E31–37.

Cheung, S.S., Reynolds, L.F., Macdonald, M.A., Tweedie, C.L., Urquhart, R.L. and Westwood, D.A. (2008) "Effects of Local and Core Body Temperature on Grip Force Modulation During Movement-Induced Load Force Fluctuations", *Eur J Appl Physiol*, 103, 59–69.

Choi, P.Y., Pope, H.G., Jr. and Olivardia, R. (2002) "Muscle Dysmorphia: A New Syndrome in Weightlifters", *Br J Sports Med*, 36, 375–6; Discussion: 377.

Denslow, S., Lomarev, M., George, M.S. and Bohning, D.E. (2005) "Cortical and Subcortical Brain Effects of Transcranial Magnetic Stimulation (TMS)-Induced Movement: An Interleaved Tms/Functional Magnetic Resonance Imaging Study", *Biol Psychiatry*, 57, 752–60.

Dumestre-Toulet, V., Cirimele, V., Ludes, B., Gromb, S. and Kintz, P. (2002) "Hair Analysis of Seven Bodybuilders for Anabolic Steroids, Ephedrine, and Clenbuterol", *J Forensic Sci*, 47, 211–14.

Federoff, H.J., Burke, R.E., Fahn, S. and Fiskum, G. (2003) "Parkinson's Disease: The Life Cycle of the Dopamine Neuron. September 18–20, 2002. Princeton, New Jersey, USA. Proceedings", *Ann NY Acad Sci*, 991, 1–360.

Friedl, K. (2005) "Effects of Testosterone and Related Andogens on Athletic Performance in Men" in Effects of Testosterone and Related Andogens on Athletic Performance in Men, Volume 9, Olympic Encyclopaedia of Sports Medicine, (International Olympic Committee, Blackwell Publishing).

Friedl, K. and Allan, J. (2003) "Physiological Research for the Warfighter—the U.S. Army Research Institute of Environmental Medicine", *U.S. Army Medical Department Journal*, 33–44.

Friedl, K. and Penetar, D. (2008) "Resilience and Survival in Extreme Environments" in Resilience and Survival in Extreme Environments, Volume 7, (Boca Raton, Florida: CRC Press, Taylor and Francis Group), 139–76.

Friedl, K.E. (2008) "Is It Possible to Monitor the Warfighter for Prediction of Performance Deterioration?" Technical Report, (Neuilly-sur-Seine Cedex, France: Workship on Operational Fatigue) RTO-HFM/WS-151, 7.1–7.10.

Friedl, K.E., Grate, S.J., Proctor, S.P., Ness, J.W., Lukey, B.J. and Kane, R.L. (2007) "Army Research Needs for Automated Neuropsychological Tests: Monitoring Soldier Health and Performance Status", *Arch Clin Neuropsychol*, 22 Suppl 1, S7–14.

Friedl, K.E. and Hoyt, R.W. (1997) "Development and Biomedical Testing of Military Operational Rations", *Annu Rev Nutr*, 17, 51–75.

Gislen, A., Dacke, M., Kroger, R.H., Abrahamsson, M., Nilsson, D.E. and Warrant, E.J. (2003) "Superior Underwater Vision in a Human Population of Sea Gypsies", *Curr Biol*, 13, 833–6.

Gislen, A., Warrant, E.J., Dacke, M. and Kroger, R.H. (2006) "Visual Training Improves Underwater Vision in Children", *Vision Res*, 46, 3443–50.

Gonzalez-Cadavid, N.F., Taylor, W.E., Yarasheski, K., Sinha-Hikim, I., Ma, K., Ezzat, S., Shen, R., Lalani, R., Asa, S., Mamita, M., Nair, G., Arver, S. and Bhasin, S. (1998) "Organization of the Human Myostatin Gene and Expression in Healthy Men and HIV-Infected Men with Muscle Wasting", *Proc Natl Acad Sci USA*, 95, 14938–43.

Haisman, M. and Vogel, J. (1986) "Physical Fitness in Armed Forces", Panel on the Defence Applications of Human and Biomedical Sciences) NATO Final Report (Panel VIII) D/125, 113.

Haller, C.A. and Benowitz, N.L. (2000) "Adverse Cardiovascular and Central Nervous System Events Associated with Dietary Supplements Containing Ephedra Alkaloids", *N Engl J Med*, 343, 1833–8.

Hirai, K., Horvath, S.M. and Weinstein, V. (1970) "Differences in the Vascular Hunting Reaction between Caucasians and Japanese", *Angiology*, 21, 502–10.

Hochachka, P.W. (2000) "Pinniped Diving Response Mechanism and Evolution: A Window on the Paradigm of Comparative Biochemistry and Physiology", *Comp Biochem Physiol A Mol Integr Physiol*, 126, 435–58.

Hochachka, P.W., Gunga, H.C. and Kirsch, K. (1998) "Our Ancestral Physiological Phenotype: An Adaptation for Hypoxia Tolerance and for Endurance Performance?" *Proc Natl Acad Sci USA*, 95, 1915–20.

Hopkins, M.G., Spina, R.J. and Ehsani, A.A. (1996) "Enhanced Beta-Adrenergic-Mediated Cardiovascular Responses in Endurance Athletes", *J Appl Physiol*, 80, 516–21.

Hultman, E., Soderlund, K., Timmons, J.A., Cederblad, G. and Greenhaff, P.L. (1996) "Muscle Creatine Loading in Men", *J Appl Physiol*, 81, 232–7.

Isacson, O. and Kordower, J.H. (2008) "Future of Cell and Gene Therapies for Parkinson's Disease", *Ann Neurol*, 64 Suppl 2, S122–38.

Kaplitt, M.G., Feigin, A., Tang, C., Fitzsimons, H.L., Mattis, P., Lawlor, P.A., Bland, R.J., Young, D., Strybing, K., Eidelberg, D. and During, M.J. (2007) "Safety and Tolerability of Gene Therapy with an Adeno-Associated Virus (AAV) Borne Gad Gene for Parkinson's Disease: An Open Label, Phase I Trial", *Lancet*, 369, 2097–2105.

Khaleeli, A.A., Levy, R.D., Edwards, R.H., McPhail, G., Mills, K.R., Round, J.M. and Betteridge, D.J. (1984) "The Neuromuscular Features of Acromegaly: A Clinical and Pathological Study", *J Neurol Neurosurg Psychiatry*, 47, 1009–15.

Killgore, W.D., Balkin, T.J. and Wesensten, N.J. (2006) "Impaired Decision Making Following 49 H of Sleep Deprivation", *J Sleep Res*, 15, 7–13.

Knapik, J.J., Jones, S.B., Darakjy, S., Hauret, K.G., Bullock, S.H., Sharp, M.A. and Jones, B.H. (2007) "Injury Rates and Injury Risk Factors among U.S. Army Wheel Vehicle Mechanics", *Mil Med*, 172, 988–96.

Kochakian, C.D. and Stettner, C.E. (1948) "Effect of Testosterone Propionate and Growth Hormone of the Weights and Composition of the Body and Organs of the Mouse", *Am J Physiol*, 155, 255–61.

McPherron, A.C. and Lee, S.J. (1997) "Double Muscling in Cattle Due to Mutations in the Myostatin Gene", *Proc Natl Acad Sci USA*, 94, 12457–61.

Nevola, V. (2009) "Common Military Task: Digging", Chapter 4, Optimizing Operational Physical Fitness, (Neuilly-sur-Seine Cedex, France:Research and Technology Organisation, North Atlantic Treaty Organisation) NATO Final Report AC/323(HFM-080)TP/200, 4.1–4.68.

Porrino, L.J., Daunais, J.B., Rogers, G.A., Hampson, R.E. and Deadwyler, S.A. (2005) "Facilitation of Task Performance and Removal of the Effects of Sleep Deprivation by an Ampakine (Cx717) in Nonhuman Primates", *PLoS Biol*, 3, e299.

Prather, I.D., Brown, D.E., North, P. and Wilson, J.R. (1995) "Clenbuterol: A Substitute for Anabolic Steroids?" *Med Sci Sports Exerc*, 27, 1118–21.

Rasmussen, N. (2008) "On Speed: The Many Lives of Amphetamines" in *On Speed: The Many Lives of Amphetamines*, (Boston: The MIT Press).

Salomon, F., Cuneo, R.C., Hesp, R. and Sonksen, P.H. (1989) "The Effects of Treatment with Recombinant Human Growth Hormone on Body Composition and Metabolism in Adults with Growth Hormone Deficiency", *N Engl J Med*, 321, 1797–1803.

Saxena, A., Sun, W., Luo, C., Myers, T.M., Koplovitz, I., Lenz, D.E. and Doctor, B.P. (2006) "Bioscavenger for Protection from Toxicity of Organophosphorus Compounds", *J Mol Neurosci*, 30, 145–8.

Schmidt-Nielsen, K., Crawford, E.C., Jr., Newsome, A.E., Rawson, K.S. and Hammel, H.T. (1967) "Metabolic Rate of Camels: Effect of Body Temperature and Dehydration", *Am J Physiol*, 212, 341–6.

Schuelke, M., Wagner, K.R., Stolz, L.E., Hubner, C., Riebel, T., Komen, W., Braun, T., Tobin, J.F. and Lee, S.J. (2004) "Myostatin Mutation Associated with Gross Muscle Hypertrophy in a Child", *N Engl J Med*, 350, 2682–8.

Spurr, G. (1986) "Physical Work Performance under Conditions of Prolonged Hypocaloria: Predicting Decrements in Military Performance Due to Inadequate Nutrition", National Academy of Science, (National Academy Press), 99–135.

Teves, M., Wright, J. and Vogel, J. (1985) "Performance on Selected Candidate Screening Test Procedures Before and After Army Basic and Advanced Individual Training", Technical Report, (Natick, Massachusetts: Army Research Institute of Environmental Medicine) ADA162805.

Treffert, D. and Christensen, D. (2006) "Inside the Mind of a Savant", *Scientific American Mind*,

Tyler, D. (1947) "The Effect of Amphetamine Sulfate and Some Barbiturates on the Fatigue Produced by Prolonged Wakefulness", *American Journal of Physiology*, 150, 253–62.

Vallerand, A.L., Jacobs, I. and Kavanagh, M.F. (1989) "Mechanism of Enhanced Cold Tolerance by an Ephedrine-Caffeine Mixture in Humans", *J Appl Physiol*, 67, 438–444.

Vital, P., Ronald, A., Wallace, G. and Happe, F. (2009) "Relationship between Special Abilities and Autistic-Like Traits in a Large Population-Based Sample of 8-Year-Olds", *Journal of Child Psychology and Psychiatry*, published ahead of print.

Warber, J.P., Patton, J.F., Tharion, W.J., Zeisel, S.H., Mello, R.P., Kemnitz, C.P. and Lieberman, H.R. (2000) "The Effects of Choline Supplementation on Physical Performance", *Int J Sport Nutr Exerc Metab*, 10, 170–81.

Yerkes, R.M. (ed.) (1920) "The New World of Science: Its Development During the War" in *The New World of Science: Its Development During the War*, (New York, NY: The Century Co.).

Chapter 16

The US Army Future Concept for the Human Dimension: Chief Human Dimension Executive Summary

Steven Chandler

Introduction

The human dimension comprises the moral, cognitive, and physical components of the Soldier and organizational development and performance essential to raise, prepare, and employ the Army in full spectrum operations. Army concepts acknowledge the Soldier as the centerpiece of the Army, but none, individually or collectively, adequately addresses the *human dimension* of future operations. This study is a precursor to a shorter concept that will join the formal family of Army concepts. It provides an integrating and forcing function that draws on other joint and Army concepts to describe those aspects of a highly nuanced human dimension interacting at all levels. Like all concepts, this study seeks to identify things that must change to meet future challenges. To do this, Army concepts first project requirements from 2015 to 2024 and describe an operational or functional problem to be solved, and then express how the future Modular Force will best operate within that set of challenges and environments. Additionally, concepts identify required future capabilities necessary to operate in the manner described in each concept.

The Operational Problem

Current trends in the global and domestic operational environments will challenge the United States' ability to maintain a future responsive, professional, All-Volunteer Force. Soldiers will operate in an era of persistent conflict amongst populations with diverse religious, ethnic, and societal values. Faced with continuous employment across the full range of military operations, the Army will require extraordinary strength in the moral, physical, and cognitive components of the human dimension. Existing accessions, personnel, and force training and education development efforts will not meet these future challenges, placing at grave risk the Army's ability to provide combatant commanders the forces and capabilities necessary to execute the National Security, National Defense, and National Military Strategies.

Solution Synopsis

The Army will need to increase its human dimension focus in both the operational Army and Generating Force in order to meet future challenges and operate in an era of persistent conflict. Improved capabilities must address the broad range of human dimension actions necessary to prepare, support, and sustain this force. The Army must maintain a proper balance of moral, physical, and cognitive development with contributions from science and technology that can enhance Soldier physical and mental performance. The Army must widen the community of practice in the human dimension to continue to explore how we can best recruit, train, and retain an all volunteer force that can operate across the entire range of military operations.

In the Beginning …

In August 2006, General Wallace, Commanding General, US Army Training and Doctrine Command directed a study be made of what he called—The Human Dimension. It consists of a thorough, credible and detailed examination of behaviors, intuition and performance (cognitive and physical) impacting decision-making and interactions with people, technologies and varying environments. "I expect the study to spur thought, motivate investigation, and illuminate, through a structured approach, a strategy for the coordinated and holistic development of future capabilities."—(General William S. Wallace)

The Human Dimension was not to be limited to combat operations, but also in the realm of force generation and the affect of shifting US demographics. It includes the analysis of the full life cycle of a Soldier from recruitment and training through deployment, redeployment and reintegration into society. The study includes families, civilians, contractors and the interrelationships of groups all within the context of the future Operating Environment.

Future Operating Environment

Current trends, globally and domestically, challenge the United States' ability to sustain a quality All-Volunteer Force in an era of persistent conflict amongst diverse religious and ethnic populations. Adversaries of the twentyfirst century will: employ terrorism, avoid force-on-force conflict, use asymmetric means, hide within urban clutter and use local population as human shields. Human Dimension capabilities will enable pre-eminence in the twentyfirst century by developing

Soldiers capable of shouldering the increased responsibility of operating in small groups in volatile, uncertain, complex and ambiguous environments.

The Operational Problem

Given the requirement for Full Spectrum Operations in an era of persistent conflict with the demands of the Army Force Generation cycle (Reset—Train/Ready—Available) and that Soldiers are the centerpiece of our formations, human capabilities are the key to winning our current and future wars. The Army must focus our Human Dimension efforts to ensure: sustained quality of the All-Volunteer Force, trained Soldiers, Civilians, Leaders and units prepared for Full Spectrum Operations (FSO), a resilient force Reset and Trained/Ready for future deployments, and prepared for complex and demanding Joint, Interagency, Intergovernmental, and Multi-national (JIIM) environments now and in the future.

What the Human Dimension Study Told Us

The concept concluded that a comprehensive approach to the Human Dimension is required. The Army must:

- Examine accessions programs, policies and entry standards and adjust retention programs;
- Change career management policies;
- Develop a holistic physical fitness program (mental, medical and nutritional health) that includes: retaining qualified physically disabled Soldiers;
- Support programs that identify and mitigate causes of stress;
- Continue its commitment to the quality of life programs for the Soldier and family;
- Exploit advances in decision-making and networked operations to support—not supplant— leadership;
- Exploit S&T enablers to enhance and augment Soldier cognitive and physical performance;
- Adapt training techniques IAW learning styles and cognitive preferences;
- Prepare Soldiers to have the self-confidence to make the right decision in tough ethical and moral situations;
- Improve cultural awareness programs, language skills and invest in cultural humanities education.

The Human Dimension Concept

The human dimension is comprised of three components—Cognitive, Physical and Social. Using available and emerging tools, the Army will:

- Optimize Soldier decision-making through enhanced screening, recurring assessment and tracking of individual's potential and attributes; develop dynamic, scalable, adaptive, immersive, sensory enabled, tailored training; develop adaptive material systems maximizing individual attributes;
- Improve Soldier fitness through comprehensive wellness programs that build aerobic/mental capacity, strength, endurance, confidence, and resilience; institute sound nutrition programs; teach stress and sleep deprivation management;
- Strengthen the Soldier's character/Warrior Ethos reflecting confidence in tough moral, culturally sensitive situations grounded in law; inculcate Army Values; educate understanding social/family dynamics, social awareness, cultural differences, respect, interpersonal relationships, and spirit; strengthen affiliation with a team, fostering cohesion, positive perception of others.

Human Dimension Outcomes

The Army Capabilities Integration Center (ARCIC) works in terms of being resourced informed, integrated and outcome oriented. In other words, capability documents reflect affordable solutions, concentrate on substance not process and bring all key elements together within the context of a Joint, Interagency, Intergovernmental and multinational environment.

Cognitive Component Outcome

A recruited Future Force (soldiers/civilians) is managed and retained based, in part, on continuous cognitive assessments (such as, Attention, Learning, Leadership, Adaptability, Decision-making, Vision). Advanced technologies and tools assist in the selection of individuals for assignments and advanced accelerated/measurable training. Enhanced training (facilitates and accelerate task learning), leader development and Battle Command systems will adapt to individual proficiencies and learning rates to maximize readiness. Commanders provided tools to best match Soldier attributes and talents to mission requirements.

Physical Component Outcome

Future Force Soldiers who adhere to a continuum of holistic fitness tailored to the individual and subsequent mission requirements (measurable physiological, neurological, psychological, nutritional, and developmental fitness training). Programs that identify, mitigate, treat and rapidly restore Soldiers who become holistically "unfit" due to combat operational and stress-related injuries. Retention of qualified physically disabled Soldiers is the norm.

Soldiers are, and will, face unprecedented physical demands contributing to combat stress. Future programs will tailor fitness and wellness efforts to the individual with the goal of establishing lifelong practices. Early areas of focus include emotional, spiritual, social, physical and financial.

Social Component Outcome

Future Force (soldiers/civilians) that functions and behaves in accordance with: law; Army values; and national/international expectations and standards. Grounded by a continuum of adaptable, scalable, and measurable training programs, inculcate appropriate actions in reaction to operational challenges in tough ethical/moral situations. Commanders with improved cultural awareness programs; Soldiers with cross-cultural learning skills; and high awareness of the linkage of ethical behavior to combat effect, Army Family values and the Army's Civilian Work Force.

Ultimately, the goal is addressing the character of the Soldier, from warrior's spirit, cross-cultural awareness and professional ethics, building a firm self-confidence to perform the mission in very difficult situations.

What We Know ...

We know that the brain is filled with over 100 billion neurons, occupies only 0.6 CFT and uses about 2 watts/hr versus a Super Computer taking 1600 SFT and using 5000 watts just for cooling. There are more synaptic connections than all known bodies in the universe and that it constitutes only 2 percent body mass, yet consumes 20 percent of the energy—alert or asleep.

We know the brain is very efficient. When learning a new task, the brain is nearly universally active. However, once it masters the task, it goes into a hybrid mode using as little energy as necessary. An excellent analogy is learning how to drive. Remember how difficult it was in the beginning? All those mirrors, pedals, steering wheel, assorted controls, traffic signs, lines on the road, other drivers, potholes and the instructor. How did we manage? Now you think nothing of it. Just jump in and take off. That's novice to expert.

Seasons of the Brain

We know that the brain has three seasons—Maturing, Adult and Aging. Up to age 25 to 30 years old, our brains mature, with the pre-frontal cortex (responsible for reasoning, executive decision-making) developing last. The brain reaches adult stage with little change until approximately age 55. In practice, we may have always known of the "weakness" of youthful leaders. For example, The Army pairs the Platoon Leader (~ 22) with a Platoon Sergeant (~ 34); a Company Commander (~ 25) with a First Sergeant (~ 37). After 55, the brain begins to cull unused cells as part of the Aging season, where one could lose from 2 percent to 8 percent brain mass every decade thereafter. Part of the loss is due to a lifetime of experience where the brain has established "templates" it can apply to new experiences. Some would call it wisdom. However, the loss can be mitigated by a lifetime of challenging your brain. For example, *"Never stop learning*."

"Men and Women ARE Different"

There are distinct differences within, and between, the sexes in how information is processed, learned and decisions made. Let's explore the first. How do men and women handle spatial navigation? Women, on the norm, prefer the use of "landmarks" whereas men prefer "Euclidian geometry." Put simply, when you use MapQuest or Google maps you are presented with two forms of information: Turn-by-turn directions on the left and a map on the right. On the norm, women prefer the turn-by-turn directions, men the map. Therefore it is not surprising to observe that women have a more difficult time learning Army navigation techniques than men, since it was designed by men for men.

Understanding the differences should directly influence future education delivery methodologies used by the Army.

Oh the Possibilities!

The Army sees Neuroscience as an enabler to maximize a Soldier's inherent cognitive potential by understanding cognitive styles, information processing and learning. Traditional tools and methods, such as psychological and physiological assessments, have been very effective. However new emerging tools and methods may provide major breakthroughs in understanding brain function and eventually lead to accurate predictive behavioral models that account for our natural variability.

Future cognitive assessment tools may:

- Predict leadership potential and decision-making capabilities—affecting training received and determine career path;
- Identify cognitive styles—to inform education methodologies;

- Identify aptitude for cross-cultural understanding for nation-building tasks;
- Identify predisposition for severe combat stress or PTSD—to take preventive care where possible and begin treatment earlier.

Future cognitive learning tools may enable:

- Creation of a cognitive gym—exercise executive decision-making capabilities;
- Creation of a cognitive UCOFT—exercise full spectrum skills;
- Accelerating learning—tailored to individual cognitive potential and preferences;
- Effective resilience-building and stress mitigation techniques.

Policy

Human Dimension is focused on providing continuing assessments of an individual's talents and attributes to: manage career path, education, maintain optimum physical and psychological fitness and provide the best possible decision-making skills. However, if policy, despite the Army's best intentions, is not sufficiently responsive to change, this effort will fail. Some policy changes should be fairly easy to change, whereas others are not. A small change in policy quickly reaps big dividends for the Army. For example:

An Infantry Soldier has a number of physical strength issues due to two deployments to Iraq. He is a sharp well-disciplined Soldier hoping to remain in the Army. Despite chain-of-command efforts to try re-classification into a less physically demanding specialty, the Army conducts a Medical Review to determine if he is fit for service in the Infantry. He was medically discharged. Allowing re-classification as part of the review board possibly retains a highly motivated Soldier and saves time and money not having to recruit and train a replacement.

Conclusion

Human Dimension is about:

- Empowering Soldiers (as individuals and small groups) to dominate the Land Domain;
- Improving cognitive, physical, and social abilities;
- Enhancing and restoring cognitive and physical performance that includes mitigating the increase in physiological and psychological stress;
- Functioning efficiently as integral component of a network and society;

- Recommending changes to policy that support emerging cognitive, physical and social tools.

> "No two persons are born exactly alike …. All things will be produced in superior quantity and quality, and with greater ease, when each man works … in accordance with his natural gifts …"
>
> (Plato, *The Republic*, c.360 BC)

Glossary of Terms

Army Capabilities Integration Center (ARCIC)—Pronounced "R-Kick", designs, develops, integrates and synchronizes force capabilities for the Army across the DOTMLPF imperatives into a Joint, Interagency, and Multinational operational environment from concept through capability development.

Army Force Generation—The Army's training and readiness strategy. It is a structured progression of increased unit readiness over time that provides periods of deployment ready units in support of regional combatant commander requirements. Every unit goes thru a three year cycle of Reset, Train/Ready and Available phases.

DOTMLPF—Doctrine, Organization, Training, Material, Leadership, Personnel and Facilities.

Full Spectrum Operations (FSO)—Army forces combine offensive, defensive, and stability or civil support operations simultaneously as part of an interdependent joint force to seize, retain and exploit the initiative to achieve decisive results. (Field Manual 3–0)

Operational Environment—The composite of the conditions, circumstances and influences affecting the employment of capabilities and bear on the decisions of the commander. It encompasses physical areas and factors (of the air, land maritime and space domains) and the information environment. Included within these are the adversary, friendly and neutral systems that are relevant to a specific joint operation. (Joint Publication 3.0)

Persistent conflict—A period of protracted confrontation among state, non-state and individual actors who continually use violence as a means of achieving their political and ideological objectives.

UCOFT—Unit Conduct Of Fire Trainer used by the Artillery to economically develop field artillery skills without actually firing expensive ordinance.

Chapter 17
Sustaining Performance in Mass Casualty Environments

Annette Sobel

Background

In an age of terrorism, the challenges posed by the mass casualty environment are all too commonplace. Tokyo, Madrid, Mumbai, Tel Aviv, New York City, New Orleans, and virtually every continent has been affected, whether through natural or man-made disasters. The mass casualty setting is particularly overwhelming due to the high probability of recurrent events and systematic targeting of large numbers of individuals, mostly civilians. Beyond coping with the psychological impact of such events, the responder community must maintain sustainable performance throughout all phases of the response and mitigation/recovery. Although personnel selection for such tasks is difficult, subjective, and has unreliable predictive value, historical performance is traditionally employed and considered most reliable of all indicators.

This paper will focus on the evolving roles of technology in sustaining performance in mass casualty environments. These environments span the tasks of mass care, early detection and recognition of the event(s), information management and triage of casualties and information, and mitigation of effects and recovery of the individual and aggregate community. The spectrum of civil-military operations that may benefit from immersive technology and application of learning management systems spans interdisciplinary platforms. Impacts may be measured in evidence-based outcomes, and are most useful when tailored to contextual and threat-based needs.

In special environments such as bio-terrorism, integration of bio-surveillance data is critical as a driver to operations and specific techniques, tactics, and procedures. In addition, mass casualty environments demand enhanced situation awareness of anomalous conditions, impose physiologic stressors due to enhanced personnel protection, and require specialized cognitive tasks above baseline. In summary, this discourse will focus on the importance of ensuring public health infrastructure and relevant technology insertion for surge capacity necessary in optimized mass casualty management.

Sustaining performance includes more global measures beyond employment of technical skills. Specifically, operational readiness and mission success include cultivating the essential characteristics of trust on individual and aggregate levels and emergence of leadership, in extremes of situations. Trust and leadership aptitude are distinctive innate traits that may emerge or be subverted due to a number of endogenous and exogenous conditions, and not wholly learned.

Triage Process

Sustaining performance throughout the mass casualty triage process is the core function to be examined. The process involves continuous re-assessment and re-allocation of resources to ensure optimal medical outcome at the individual and aggregate levels. In a civil-military environment such as humanitarian assistance/disaster response, managing casualties with extremes in age adds significantly to the complexity of this process. This author will describe the triage process using the metaphor of the skin of an onion.

The outer skin of the process is the assessment of threat, and rapid implementation of adequate security measures. Next is the iterative feedback process of resource allocation coupled to casualty stabilization and survivability assessment. In a military setting, some of the societal norms that apply to the triage process are adapted to fit the situation. For example, patient privacy and application of resources to the elderly must be a judgment call based on the "good of the whole" and need for expedient care.

Assessment Methodology

The primary methodology of assessment of categories of triage in an unconventional threat environment is underpinned by threat agent assessment. The standard taxonomy for biological agent assessment is categorized A, B, or C, depending upon pathogenicity, transmissibility, and survivability. The primary assessment of mass casualty survivability and subsequent distribution of resources follows conventional management procedures followed by a series of unique considerations. These considerations follow a loose taxonomy of naturally-occurring/man-made biological agents, chemical agents, industrial chemicals, toxins (possessing characteristics of both biological and chemical agents), and next generation agents. The latter category is beyond the discussion of this paper, but deserves note due to the potentially confounding signs and symptoms within the context of a mass casualty environment.

Human performance depends upon an appreciation for the environmental conditions and individual susceptibilities of casualties and response personnel, emerging and evolving threat parameters, the range of physiologic responses associated with trauma with/without threat overlay, and other tactical or operational considerations (that is, response time, geographic and temporal dispersal, logistics support, communications, and so on).

A systems engineering approach to mass casualty management is preferred. Such an approach is based upon integration of the major, highly interactive components of early recognition and characterization of the event(s) and early decisions regarding rules of engagement for distribution of resources. Sustainability of performance is key in such a model.

Key observable performance parameters include: time to task completion; accuracy of task completion; appropriateness of decision-making algorithm; appropriateness and timeliness of data integration; time to casualty reassessment. Resulting measurable outcomes include: probability of patient morbidity and mortality; probability of aggregate morbidity and mortality (casualties/responder); probability of detection, identification, and categorization (for example, threat, patient condition, friendly forces, and so on); sustainability of consistent performance; effective communication; effectiveness and timeliness of skills/procedural performance.

Initial baseline measurement of performance and effectiveness, with and without a number of environmental, physiologic and decision-making stressors should be completed. A family of performance curves will result. These curves should be validated as best as possible with simulated conditions, pilot studies, historical data, or a combination as appropriate. The analyst or decision-/policy-maker should recognize that context is everything in mass casualty management and acceptable performance parameters in a combat environment may vary greatly from humanitarian or disaster relief environments.

Subsequent determination of acceptable upper and lower "thresholds" for mass casualty performance parameters varies widely in unconventional versus conventional threat environments. For example, unconventional threats may be more sudden in onset, time compressed and of greater intensity than naturally occurring or conventional threats. In addition, the characteristics of the traumatic injuries to include severity and geographic distribution may be much greater in an unconventional threat environment. Threats may be recurrent or unpredictable in nature or timing in either environment, and the recovery and response periods vary widely. Each scenario is traumatic to responders, observers, and casualties alike, with psychological aftermath, and secondary and tertiary effects. These effects are as far-reaching as diminished confidence level in government pre-/trans-/post-event effectiveness and ability to sustain and provide basic services.

Alternatives in Mass Casualty Incident (MCI) Education and Training

Problem-based learning is the preferred approach to MCI education and training. Complex environments demand the understanding of multiple, dynamic systems, and the ability to rapidly synthesize information and determine what is important. Triage algorithms form the straw-man decision-assisting structure for such environments. Application of these algorithms in the framework of a problem-based approach with contextual relevance forms the basis of effective education and training. Measures of performance and effectiveness may be derived from such a framework.

Optimal performance results from complementary and synergistic interaction of the student, instructor, and the training/educational platform. An ideal system is adaptive and includes design elements recommended by subject matter experts. Since the effectiveness of mass casualty management relies heavily on successful performance of non-medical tasks such as communication, logistics, security, reconnaissance and threat/vulnerability analysis, surveillance, cultural appropriateness of tasks, and data and knowledge management, the establishment of acceptable performance norms is not only challenging but also dynamic. This author has experienced that a baseline acceptable performance level must be pre-determined and subsequently readjusted to accommodate the specific environment and unique characteristics of the mass casualty event. In order to ensure completeness and contextual accuracy, performance analysis should be performed pre-, trans-, and post-incident(s), with recognition that there is rarely clear distinction between phases of an attack, rather these phases present as a continuum. When feasible, each phase of the assessment should include essential task management (checking vital signs, stability, and others), critical decision-making under stress, team management skills (crew resource management and communication), and iterative situation assessment and dynamic readjustment of action plan, based upon incorporation of new or altered information.

In order for performance metrics to be transferable from a training or educational environment to a real-world environment, several qualities are germane to the design and implementation of the environment. These qualities involve complex interactions and possible overlap and include:

1. realism of the scenario, tasks, sensory input systems;
2. real-time or near real-time performance feedback;
3. transparency of the simulation environment that results in "believability" and appropriate levels of physiologic stress response;
4. appropriate, timely, and physics-based interaction of key entities to include medical and non-medical responders, medical professionals and casualties, the micro- (including instruments, patient, lighting, and so on) and macro-environment (such as infrastructure and weather) and action taken.

Mass Casualty Incident Scenarios

This section will describe the general characteristic of biological agents making them attractive to terrorists, basic triage and decision-making algorithms essential to achieving acceptable MCI performance and outcomes in the near (hours to days) and short (days to months) terms. The selected scenarios attempt to capture the spectrum of human performance likely to be encountered in these high consequence events. We will assume a single event and a defined geographic region with high certainty and confidence that the event is real.

Biological agents may be considered "ideal terrorist weapons" due to a number of characteristics. Most notable among these are ease of access, dual use, and natural occurrence, emergence and re-emergence of pathogens of interest. Furthermore, biologicals intended for legitimate use may be obtained from supply houses and research laboratories and subsequently weaponized. Technologies and infrastructure, such as those employed for fermentation, antibiotic and vaccine production may be readily modified and scaled for required production purposes. Ease of scalability, deployment, and concealment in the natural environment is described in the Interpol bio-terrorism fact sheet, http://www.interpol.int.Public/BioTerrorism. Unfortunately, many biological agents are spread with relative ease through air, food, and water. Most media produce dilutional effects, and water treatment systems further reduce pathogenicity and transmissibility of many agents. In contrast, in humanitarian assistance and disaster response settings, human-human, and human-animal proximity and population co-location with untreated water sources make such settings much more vulnerable to disease outbreak, and necessitate rapid detection, response, and proactive measures.

The fact that approximately 85 to 90 percent of biological agents are naturally occurring and zoonotic (for example, transmitted between animals and humans) compounds the above challenges. Although more difficult to detect, a deliberate disease outbreak may be distinguished from one that is naturally occurring based on the context of the outbreak. Specifically, contextual information such as temporal-spatial distribution and seasonality of the outbreak may be inconsistent with a naturally occurring event. The above points illustrate the necessity of continuous situational awareness, iterative data sampling, and anomaly detection are essential components to optimal human performance in an unconventional event(s), and are critical to survivability.

Illustrative Scenarios

1. Humanitarian assistance with emerging infectious disease(s), both respiratory and water-borne. (Flooding in a remote region with large displaced, transient population.) Significant measures of performance and effectiveness include: immediate implementation of security measures; use of proactive public health measures; threat assessment and containment; cultural evaluation; patient registration to include identification and sorting in family units; establishment of potable water sources; resource inventory.
2. Terrorism event with blast injuries and chemical contamination (Train derailment with toxic payload in an urban environment). In this scenario, significant measures of performance and effectiveness include: immediate threat assessment and threat containment; protection of casualties and

responder personnel from harm; casualty evacuation; meteorological assessment; accurate and timely threat detection and identification; resource inventory; iterative reassessment.

Summary

Human performance assessments within training environments have limited utility in prediction of behavior and adaptation in real-world conditions. Although we strive for the admirable objective of sustaining performance under stress, the metrics used are derived from conventional standards in settings where linear reasoning prevails, and outcomes-based medicine is the normative. The multi-factorial, stressful conditions encountered in MCIs are extraordinarily difficult to emulate, and most importantly, performance cannot be extrapolated with a high degree of certainty from training environments. By definition, MCIs are complex emergencies with many moving parts and rapidly changing, highly interactive determinants of outcome. Baseline assessment of the environment and determination of critical elements of information are considered the highest priority tasks to be performed by leadership and responders alike. In conclusion, a systems engineering framework emphasizing interactions and courses of action most likely to succeed models the most pragmatic approach to problem-based, contextual learning.

Design, test, and evaluation of training and educational platforms are essential to ensure optimal opportunities for performance enhancement and transferability of the training environment to the real world. Although such platforms may emulate wildly disparate man-made and natural events with unique characteristics of mission success, basic decision-making skills and reassessment tools and techniques remain universal. For example, prioritization of treatment and anticipated resource allocation based upon probability of survival and intensity of essential services needed is required. However, in the unconventional threat environment, a number of unknowns compound the decision-making process and introduce error to management when conventional norms are applied. Hence, rules-of-engagement for mass casualty event management must be clearly and unequivocally established and adhered to, and combined with continuous situational reassessment.

The disparate scenarios presented in this paper are representative of realistic unconventional, deterministic situations. Although the superimposition of biological and chemical effects is unlikely, and conventional, mega-terror events most likely due to explosives, the 1995 Sarin gas attacks on the Tokyo subway and the 2001 Anthrax letter attacks taught the world that mass fear, panic, anxiety, general disruption to society and post-traumatic stress disorders may be attenuated through knowledge, iterative training and emphasis on situation awareness with evidence-based validation. These events generated large-scale cooperative behavior, recovery of society and return to a generally higher level of awareness and functionality. Rebound of public confidence in government services, reliance of all sectors of society on media for decision-assistance, and return of rule-of-law remain reliable indicators of effective mass casualty management.

In summary, the human dimension of sustaining performance remains the most challenging. Immersive training and educational environments represent one approach to reinforcing a cascade of tools intended to encourage autonomous but interactive decision-making and learning. These environments have demonstrated added value in reducing the clutter from information overload, promoting anomaly detection in a rapidly changing, and highly interactive environment, and accelerating the emergence of the "expert" student. Selection of performance metrics will ideally reflect a systems-based approach to the individual and team interactions necessary for scalability (such as surge capacity), survivability and sustainability of performance.

Chapter 18
Neuropsychiatry and Aviation Safety

Robert R. Ireland

Commercial aviation accident rates have progressively dropped to levels so low that safety communities may be challenged to readily identify residual areas for improvement. Such was the conclusion of the Federal Aviation Administration's 2006 publication, "Human Error and Commercial Aviation Accidents: A Comprehensive, Fine-Grained Analysis Using the Human Factors Analysis and Classification System (HFACS)" (www.faa.gov/library/reports/medical/oamtechreports/2000s/media/200618.pdf) as written by Shappell and colleagues. In that same year, the United States Air Force announced its safest year in aviation with eight destroyed aircraft and one aviation fatality, in contrast to over 500 aircraft destroyed at the cost of 500 lives in 1947.

Even if it is possible to reduce accident rates further, such would require a particular constellation of circumstances in order to demonstrate. Ever more subtle, less obvious mishap causes (or roots of known risk phenomena) would need to be identified, thus including factors falling already within the rubric of the HFACS. Due to the rarity of fatal accidents, a long period of time would be required to demonstrate whether a process to reduce their incidence actually does.

To demonstrate their relevance statistically, such factors would require applicability to a broad range of flight operations and involve highly prevalent factors. To justify pursuing such factors further, those selected for study should ideally lend themselves to manipulation with practical interventions that do not pose additional risks themselves.

Due to improvements in airframes and avionics, the frequency of accidents associated with aircrew or supervisory human error historically averages 70 percent. While the physical environment is a pre-condition for unsafe acts in 58 percent of fatal commercial aviation accidents, unsafe acts of aviators are also highly associated. These include skill-based errors (56.5 percent), decision errors (36.7 percent), perceptual errors (6.5 percent) and procedural violations (23.1 percent).

Common skill-based errors involve monitoring airspeed and aircraft control. Common decision errors involve in-flight planning and decision-making. Common violations include intentional flying under visual flight rules (VFR) into instrument meteorological conditions (IMC), as well as procedural and directive violations. While the 2006 FAA analysis of commercial aviation accidents offers no major surprises in these areas, its quantification of the impact of specific human factors provides measureable targets for approaches to mitigating them.

From a neuropsychiatric viewpoint, mishap errors and violations may reflect levels of cognitive impairment in areas of attention, memory and executive functions such as initiation, planning, execution and anticipation. Executive functions control attention, inhibition, set-shifting and task management. These particular domains can be correlated to specific mishap human factors. Further, corresponding brain anatomy, neurocircuits, and neurotransmitters can be further correlated to these factors.

Several non-pathologic conditions influence behavior, including largely heritable temperament, environmentally influenced character development, unhealthy lifestyles, and external factors such as losses as grief or divorce. Modifiable pathological conditions that may be related to mishap human factors including medical conditions (as endocrine dysfunction, brain lesions), physiologic events related to flight as hypoxia, and more commonly, mental disorders.

Some mental disorders, such as major depression, are also associated with cognitive impairment in a fashion similar to that resulting from moderate head injury. Mild dysphoria to full-blown major depression result in alterations of cognition directly related to human factors commonly associated with aviation mishaps. These include deficits in: attention (as to air speed and aircraft control); memory (resulting in skill- and decision-based errors and violations); and executive functioning (focusing attention, set-shifting, task management, as well as dysinhibition (and violations).

Thus, improved identification of even relatively mild mental disorders such as depression in aviators may be "low hanging fruit" in aviation safety in that they are highly prevalent and do affect cognitive domains related to usual human factors in mishaps. Such disorders are also treatable. Current limitations to addressing mental disorders in aviators relate primarily to policies that tend to be overly punitive, and those of most aeromedical authorities that limit maintenance treatment options after return to piloting aircraft when disorders are in remission.

The study of the results of recent policies that actively encourage aviators to treat such disorders and allow the controlled use of psychotropics to prevent recurrence of symptoms will inform aeromedical authorities still using traditional and more restrictive approaches. Aeromedical authorities with policies allowing treatment of aviators with maintenance medication include Canada (civil and military), Australia (civil only), and the United States (Army only). The degree to which aviators fly with mental disorders is difficult to measure, especially when penalties for disclosing such conditions are perceived as career threatening. Reducing such threats through monitoring, and moderating policy that requires permanent career termination if a second episode of a condition such as major depression recurs will (like recent alterations in alcohol abuse policies) encourage aviators with impairing conditions to more frequently come forward for evaluation and treatment.

References

Shappell, S., Detwiler, C., Holcom, K., Hackworth, C., Bouquet, A., Wiegman D. (2006). "Human Error and Commercial Aviation Accidents: A Comprehensive, Fine-Grained Analysis Using HFACS-Human Factors Analysis and Classification System." DOT/FAA/AM-06/18, Office of Aerospace Medicine, Washington, DC, 20591

Chapter 19
Closing Remarks: Realization of Existing Capabilities in Sustaining Performance

Rebecca M. Steinberg

Michael D. Matthews

Steve Kornguth

Introduction

US soldiers on the modern battlefield are engaged in extended operations of an indefinite duration and unprecedented complexity. In conditions of asymmetric warfare, troops contend with enemies embedded in a civilian population. To ensure survival and mission success in this environment, soldiers cope with substantial cognitive challenges that must be addressed rapidly within the threat environment (seconds to minutes). These cognitive demands include maintaining active vigilance of the surroundings, assimilating networked data transmitted to multiple devices and operating platforms, formulating appropriate and timely plans of action, and communicating with peers and commanding officers. While these requirements alone are sufficient to generate stress, soldiers are additionally affected by extreme climate, sleep deprivation, cultural dissonance in foreign territories, physical fatigue, drug/pharmaceutical use and abuse, prolonged separation from family, and threat to physical wellbeing. Soldiers are highly trained and capable, but current knowledge can be used to improve their abilities even further during periods of stress and fatigue.

The authors contributing to this volume were presenters at the *Sustaining Performance Under Stress Symposium*, held on February 26–28 in Adelphi, Maryland. The conference was organized to meet the need for increased cross-communication between researchers, government officials, and military leadership regarding capabilities and directions for mitigation of impaired performance during stress. The contributors to this book present their expert opinions on the neurophysiological correlates of stress, methods for measuring or modulating stress responses, and guidance for the transition of current knowledge to tangible benefits for the soldier. Several contributors with firsthand experience of stressors in the combat zone have shared their vision of directions for future research. The goal is to generate data, devices, and therapeutics that can be put to use for soldiers on the battlefield within a 5- to 10-year time span. Related to this goal is the desire to pre-identify individuals who are likely to exhibit susceptibility or resilience to the negative effects of combat stress. Once identified, such individuals could be closely monitored and provided with adequate resources to confront stress-related deficiencies at their onset.

The necessity of combining science with procedural needs, highlights a cultural difference between scientists and military professionals that must be overcome. Scientists are conservative in data interpretation and prefer stepwise and logical progression of research topics, while the military rapidly adjusts its priorities to meet the demands of a highly dynamic battlefield. The military would benefit from knowing 70 percent of the answer to a crucial question in the near-term rather than achieving a 100 percent understanding in the long-term. Science, however, progresses slowly to yield more thorough data, but in a dated timeframe that diminishes the utility of the information to the military. Mutual recognition of these differences would permit both sides to work more closely towards a common goal.

The chapters in this volume fall into four sections. The first four chapters use real-world activities to investigate cognitive burdens and performance (Genik, Cummings and Nehme, Rizzo and Severson) and explore the utility of neuroimaging devices for real-world activities (Dunn). The next four chapters address sleep deprivation effects on cognition and performance (Rocklage et al., Dinges and Goel, and Stickgold), and pharmaceutical countermeasures (Hampson and Deadwyler). Following these are three chapters that explore cognition during stress and anxiety, including emotional processing (Paulus et al.), intuitive thinking (Kounios), and decision-making (Liberzon and Gonzalez). Hatfield, Merlo, and Friedl discuss neuroergonomic aids: their benefits and potential pitfalls. Finally, Martz and Chandler, Sobel, and Ireland emphasize the primary importance of soldier cognition to maximize Army capabilities

Cognition During Real-World Activities

Stressful situations endangering life and limb such as those experienced in battle cannot and should not be recreated in a controlled laboratory setting. Milder stresses such as those associated with cognitive burden during complex tasks like driving, lend themselves more easily to laboratory study. Genik translates complex tasks such as driving to computerized scenarios played on a monitor while volunteers undergo functional magnetic resonance imaging (fMRI) or magnetoencephalography (MEG). The controlled laboratory setting provides the benefit of highly precise measurements of brain activity while the test driving scenarios are directly applicable to real-life stressors. This paradigm has established which brain regions are likely to contribute to the cognitive processes associated with concentration/distraction during driving. The near-term capabilities enabled by this research include the development of a state-descriptive model using activity in the right superior parietal lobe to identify when the driver's concentration falters.

Cummings and Nehme test behavioral models that predict optimal performance based on workload, focusing on the military-relevant task of remote control of unmanned vehicles. When incorporating the classic Yerkes-Dodson inverted U-shaped curve of the workload versus performance function, the model best described actual human performance on the task. This predictive human performance model can be used to design work systems that autonomously parcel workload responsibilities amongst operators to generate maximum efficiency. It would also be informative to test such models on brain activity in *a priori* selected brain regions related to cognition and attention.

Rizzo and Severson employ standardized statistical methods and a variety of data measurement techniques to understand how people behave in response to events that they actually encounter in their quotidian lives. They investigate electroencephalographic (EEG) brain activity, location (global positioning system), movement (accelerometer), and heart rate during complex real-world experiences, including driving. Investigations into the effects of real-life stressors on neurophysiology and performance can reveal the difference between extreme and mild stress on performance.

Dunn examines the utility of near infrared spectroscopy (NIRS) as a portable, non-invasive measure for changes in brain activity during real-world events. With refinement of the equipment and analysis measures in the next 5–10 years, it may be possible to generate a lightweight portable NIRS device for recording brain activity in the field. NIRS represents an advantage over other neuroimaging systems in that it employs harmless light waves to sense changes in blood oxygen level, the same metric observed in fMRI. For this to be effective, NIRS data should be correlated with fMRI measures during specific particular tasks.

Cognition During Sleep Deprivation

Several of the Authors Focus on the Effects of Sleep Deprivation on Cognition.

Some have developed methods for pre-identifying soldiers who are susceptible or resilient to sleep deprivation. Their methods include analysis of white matter tract fractional anisotropy (Rocklage et al.), genetic testing for alleles with known relationship to sleep deprivation susceptibility (Dinges and Goel), and standardized behavioral testing during sleep deprivation (Stickgold, Dinges, and Rocklage et al.). Their research identifies susceptibility to sleep deprivation in groups of people, a necessary step prior to predicting performance in specific individuals. It is probable that no single measure will predict with high reliability whether an individual will be resilient or susceptible to sleep deprivation. Rather, a combination of metrics will likely be necessary.

Hampson reports the relative efficacy of various pharmaceutical aids in sustaining cognitive abilities during sleep deprivation, using non-human primates. The noninvasively administered pharmaceuticals target the excitatory glutamatergic neurotransmitter system as well as a hormone system (orexin) recently discovered to mediate sleep and wake states. While novel cognitive aids to improve decision-making during fatigue without increasing sleep-debt would be a valuable tool for military and industrial use, the relative risks of addiction and overdose should be assessed. It will be interesting to determine whether specific pharmaceutical therapies can be targeted for different individuals. For example, an individual possessing a particular genetic polymorphism conferring impaired performance during sleep deprivation may respond differently to pharmaceutical interventions than someone with a different predisposing trait.

Different abilities decay at different rates following sleep deprivation. Whereas total sleep time is a primary factor in determining performance on a psychomotor vigilance task (Dinges and Goel), time spent in particular stages of sleep is a more important factor in memory formation and recall (Stickgold). Stickgold notes that skill on motor learning and visual perception tasks improves following slow wave sleep occurring either during overnight rest or during a 60–90 minute nap. Emotional memory formation, however, is reliant upon rapid eye movement (REM) sleep. Sleep deprivation specifically augments memories of negative emotional images while impairing memories of positive emotional images, and this trend can only be reversed through REM sleep. The implication for soldiers in the field is that sleep deprivation may enhance the remembrance of negative visual images, which could contribute to battlefield stress. Knowing which abilities are affected by different aspects of sleep, and the timeline of decay for particular cognitive skills, would permit a better assessment of soldier readiness for different tasks. Increased variability in performance can also result from sleep deprivation (Dinges and Goel), which could decrease reliability and impair team dynamics.

It may be possible that decreased cognition during sleep deprivation is a function of interrupted concentration due to frequent microsleeps (brief<30 second unintentional sleep bouts), rather than to a general decline in cognitive abilities. As a corollary question, it would be interesting to know whether pharmaceutical aids to improve awareness during sleep deprivation impact microsleep number or general brain arousal. The role of microsleeps in cognition during sleep deprivation should be further investigated.

Cognition During Stress and Anxiety

Stress and anxiety can disrupt or potentiate cognitive skills, depending on the situation and the individual. Three contributors to this book investigate the interplay between cognition and stress/anxiety for psychosocial tasks, decision-making, and intuitive thought. Paulus et al.'s research has determined that individuals deemed *a priori* to be resilient to stress and adversity differ in activation of the ventromedial prefrontal cortex and amygdala in response to emotional faces, compared with individuals judged to be susceptible to stress. It would be interesting to evaluate whether those individuals resilient to stress according to Paulus' criteria are similarly resilient to the effects of sleep deprivation for the same emotional processing task. This would reveal whether abnormal brain activity in these regions correlates with susceptibility to post-traumatic stress disorder (PTSD).

Intuition is defined as the sudden solution to a seemingly intractable problem, as contrasted to a step-wise deduction/induction model. Intuitive decision-making is widely, albeit anecdotally, associated with excellence in leadership, particularly during times of impending danger (high stress). Until a relatively recent refinement in the resolution of neuroimaging, it has been impossible to determine which brain networks underlie intuitive decision-making. These traits would be valuable on the battlefield by decreasing reaction time and increasing innovation against an enemy that plans offensives based on predictable and habitual activities of our forces. Kounios provides preliminary evidence for the neural underpinnings of intuitive decision-making. If individuals with greater intuitive decision-making abilities could be identified at enlistment, it would represent a major benefit to the Army.

Liberzon and Gonzalez promote research to separate out the relative contribution of cortisol from the effects of psychological stress during decision-making. Their chapter suggests that treating participants with the stress hormone cortisol during a stress-inducing behavioral task renders the participants more attentive to the magnitude of a reward but less sensitive to the probability of winning the reward. They remark that there may be two distinct neural networks guiding assessment of reward versus probability. If so, slight manipulations of cortisol might improve soldier decision-making under stress.

Neuroergonomic Aids

Certain technologies are currently being evaluated for utility on the field in sustaining cognitive abilities and brain activity. One such device utilizes tactile senses to improve message transmission. Lieutenant Colonel Merlo has innovated a vibrational signaling belt to aid in communications in visually obscured environments (darkness, smoke, or dust), and/or during stress and sleep deprivation. This elegant approach to improving performance should be thoroughly assessed to determine whether the tactile signal might interact constructively or destructively with visual and auditory communications, especially during high operation tempo activities and in stressful encounters.

Much interest has been focused lately on brain training devices—learning methodologies intended to improve overall cognitive function. While few of these applications have been objectively assessed using scientific methodology, Hatfield provides compelling evidence that targeted neurofeedback training promotes brain activity associated with excellent performance and improves scores in expert sniper shooting. While many commercial companies attribute similar advantages to their brain-training products, only those findings showing statistically significant improvement on an occupationally relevant task in a controlled experiment should be considered for follow-up investigations and eventual adaptation to military training protocols.

Current scientific knowledge does not enable enhancing human performance to a superphysiological state. This is likely to be possible in the near future with pharmaceutical aids such as Adderall or Ritalin, and neuroergonomic devices like the combat exoskeleton. These performance modifiers should be tested with caution in order to detect any unintended and potentially negative consequences. Some clear examples from history are mentioned by Friedl, including the dispersal of amphetamines to US soldiers in World War II, which boosted awareness and mood but also increased risky and dangerous behaviors. It is furthermore possible that successful acquisition of super-human abilities in one realm can impinge on abilities in other realms. Continuing research into brain networks underlying cognitive performance during stress with and without the neuroergonomic aid can help to identify any issues at an early stage of development.

Guidance from Military Leadership

Several military leaders shared their understanding of soldier needs and the situational demands of war. General Martz and Colonel Chandler in their contributed chapter discuss the US Army's (Training and Doctrine Command) conceptualization of how research and development can address Army needs and improve the performance of soldiers. This information is freely available as a publication entitled, "The US Army Study of the Human Dimension in the Future 2015–2024," that can be found online at http://www.tradoc.army.mil/tpubs/pamndx.htm. The Human Dimension report firmly situates the Soldier as the central focus of Army capabilities.

In her chapter, Major General (retired) Sobel stresses the available means to predict, assess, and improve the performance of first responders to a mass casualty incident. When assigning individuals to a team, benefits can be gleaned from taking into consideration the strengths and abilities of different individuals, so that team effectiveness is enriched by the contributions of selectively matched members. The same principle could match instructors with particular teams to maximize the effectiveness of training. Sobel supports the alignment of scientific with military goals by stressing that science must be used to adequately assess optimal performance. The end-user (military professionals) must direct the design of supportive technologies and techniques.

Military leaders are striving to improve the working environment and the abilities of their soldiers. For example, the adaptation of US Army doctrine to soldiers' mental health is described by Colonel Ireland, who documents efforts by military leaders to reduce air flight incidents due to human factors, while providing for the psychological needs of the soldier under stress. Oversight of the cognitive strengths of flight operators and pilots is crucial, given the high consequences of error (Buffalo, NY crash February 2009; San Diego, CA crash December 08; New York City, NY emergency landing January 2009).

Military personnel have skill sets and situation awareness that differs from the general population. The military cohort has gone through specialized instructive training and disciplined physical exertion, exists in a distinct social and cultural structure, and may have experienced extreme trials of abilities and intense emotions in the context of threats to self and peers. Scientific studies should seek appropriate control cohorts with this factor in mind.

Future Directions for Sustaining Performance Research and Development

Non-invasive portable sensors that can be facilely incorporated into the equipment and routine of the soldier are needed to capture physiological events occurring during unpredictable stress. The data from these sensors could be tested in a hypothesis-driven manner to investigate the physiological mechanisms underlying the quality of performance during stress. Such data could also be correlated with long-term changes in psychological health resulting from battlefield stress in order to develop countermeasures and avert negative outcomes.

Sensor data could be standardized for real-time monitoring of troops. Autonomous data fusion and analysis would be required to enable direct transmission of crucial and meaningful information regarding soldier status to a leader able to address the emergence of a pathological state. This information should be displayed in an easily comprehensible format such as simple color icons of red (<20 percent capability), amber (20–70 percent capability), and green (>70 percent capability). This proposed display technology incorporating a range of soldiers' physiological data is referred to as the "Commander's Dashboard." Figure 19.1 gives a schematic representation of the Commander's Dashboard concept showing how multiple measures will be fused and then individual data will be combined across an organizational unit to return a single value representing summed troop readiness.

The output will summarize a daily history of unit preparedness as a percentage of operational readiness based on the summed metrics of changes in cognitive capacity, sleep history, disease, prescription drug use, hormone levels reflective of physiological stress, and basic metrics such as heart and breathing rate. If the user requires additional information, the icon could be clicked to access more detailed group and individual data points. These metrics should be recorded by the sensors at unpredictable times and the data transmitted infrequently in short pulses in order to discourage eavesdropping by enemy forces. Such a system must not add significantly to the weight the soldier must bear and is limited by current battery technologies. We anticipate progress in alternative energy over the next 5–10 years such that lightweight renewable energy sources will be available to power remote physiological sensors that could be placed on many (perhaps eventually all) soldiers. The real-time status updates of troop preparedness and vital signs will aid the commanding officer in planning maneuvers, assigning sleep schedules, logging the cognitive and physical health of the soldiers, and identifying downed or injured individuals on the field.

Current instruments for neuroimaging can be augmented through research and development to improve their utility in the field and for measuring group interactions (rather than individuals). While fMRI is a powerful imaging device, it requires a signal averaging across several minutes of repeated stimulus presentation in order to generate a signal that is stronger than background noise. fMRI recording is also an uncomfortable, loud experience that requires one to lie prone and still for long periods of time, a situation in which it may be argued that participants might not exhibit typical behaviors. EEG recordings are taken in real-time (no signal averaging), but even slight movements of the head, jaw, and eyelids can perturb the recording, making it difficult for autonomous analysis to distinguish true brain activity from movement artifact. EEG or NIRS devices that can automatically account for movement artifacts will further facilitate studies of cognition in freely interacting individuals, or in soldiers working together as a team.

Because baseline values for many parameters vary between individuals and change with circadian rhythms, each soldier's physiological data should be assessed in comparison with his/her previous measurements. The preferred method would be to measure endpoints within each individual over time, with special care to record baseline data prior to training and prior to deployment. Preliminary studies will determine a limited number of physiological endpoints correlating reliably with cognitive state, fatigue state, physical abilities, stress-levels, and state of physical injury. Identifying a small, discrete set of variables upon which to focus research and development and minimize testing times, will render such tests feasible given the cost and time constraints of the military.

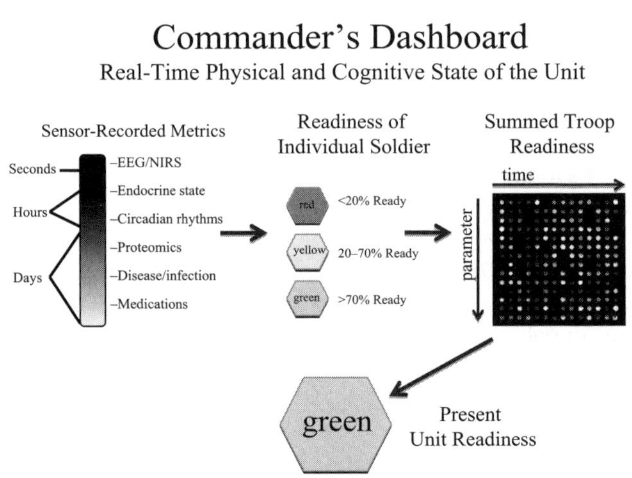

Figure 19.1 Commander's dashboard schematic

Table 19.1 identifies those areas of military criticality that the authors believe will be amenable to rapid progress in the next five to ten years based upon technology shown in the right hand column. While nutritional amendments and pharmaceutical aids can extend vigilance during long periods of time-on-task, they may also necessitate longer rest periods between missions. Item (a) in Table 19.1 addresses which technologies may assist in measuring additional time requirements to reset vigilance. Item (b) alludes to the measurement of time required to train a soldier to a new task where the operator will respond appropriately to a challenge more than 90 percent of the time. Item (c) addresses the question of which technologies can assess soldiers' cognitive flexibility, defined as the ability to adjust planned strategies rapidly in response to the threat environment. This last item refers to the technologies that may facilitate a soldier's performance as a member of a team rather than as an individual operator. Item (d) is critical because most of our neuroscience technologies are applied to the individual, while almost all operational situations in the Army require team/socialization responses.

Evidence provided in this book and elsewhere suggests that certain innate factors and life experiences can impact a soldier's ability to maintain functionality during intense stress. These factors fall into two general categories: stable traits and changeable traits. Stable traits do not change over time. For example, heritable DNA code is unchangeable without risky interventions such as gene therapy. Certain mutations or polymorphisms in the genetic code can modify overall performance by affecting neuronal signal transmission, altering drug and toxin metabolism, or impacting neuromuscular coordination or cardiovascular efficiency during stress and fatigue. Although the DNA code is immutable, the downstream effects of different gene polymorphisms can still be augmented through neuroergonomic and pharmaceutical interventions. Changeable factors relating to stress reactivity include baseline and event-reactive hormone release. A classic example of this is baseline (morning) serum cortisol, which can be altered by psychological stress, poor diet, and sleep deprivation. Altered cortisol levels can in turn impact immunity to diseases, reactivity to further stressors, and cardiovascular metrics.

Table 19.1 Critical research areas for benefit to the military within 5–10 years

Investment Areas with High Potential Payoff	Technologies
(a) Maintain Performance and Determine Recovery Interval Following Sleep Deprivation	Genomics (a, b, c) Proteomics (a, c, d) Neuroimaging (a, b, c) (such as fMRI, DTI, EEG) Pharmacology (a, c, d) Endocrine Markers (a, c)
(b) Determine Time Required to Train to Criterion	
(c) Assess Cognitive Flexibility	
(d) Measure Team Trust	

The burgeoning field of epigenetics is likely to provide insight into the stress response. Epigenetic modifier proteins either lock-down or upregulate the expression of specific genes in response to particular life experiences. These molecular changes can last for days to years, altering behavioral responses to learning and memory, circadian rhythms (including sleep), and stress. It may be possible to measure epigenetic changes in the brain by examining a proxy. For example, in animal models circulating immune cells (lymphocytes) show parallel epigenetic changes in response to stress as cells in the hippocampus (a brain region involved in learning and memory). In the future, epigenetics may reveal novel information about how and why PTSD endures for so long following a stressful event.

Here, we identify four specific areas of potential gain that are achievable in the 5-year time frame. Rather than an exhaustive list, Table 19.2 presents a select few areas that may be highly productive for the armed services in the near-term.

The first item in the above table refers to the Commander's Dashboard discussed previously. This device is currently attainable through existing knowledge and technologies. To achieve this, it is necessary to quantify the relative contributions of different metrics in determining overall Force readiness. Also, logistical concerns (sensor power source and positioning on the body, data analysis and networked data transmission) must be evaluated prior to deployment of this capability. We see this goal as a major focus for the military, returning a potentially great capability.

Table 19.2 Specific topics providing greatest deliverable benefit to the military within 5 years

	Research	**Potential Military Benefit**
	Prototype "Commander's Dashboard" testing, using metrics of EEG, heart rate, accelerometer, using algorithms	Real-time force readiness measures
Low-Hanging Fruit (5 years-attainable)	Teach intuitive decision-making via the development and testing of "brain-training" software	More rapid and innovative decision-making for asymmetric warfare
	Develop a test for serum hormone levels using sweat-patches (for cortisol, and others)	Rapid, noninvasive, inexpensive, disposable measure of stress levels
	Correlate EEG, eye-tracking, and NIRS to fMRI-observed brain activity during concentration/loss of concentration	Deployable light-weight sensors for real-time monitoring

The second item in the table refers to training to improve decision-making abilities of soldiers in the field. This is a high priority in the current state of warfare because the armed forces recognize that improved cognition is key to winning the war against terrorism (see the Human Dimension report). Rapid and intuitive decision-making might be a transferable skill shared through use of "brain-training" software, thereby enabling more expedient and accurate decisions, particularly during periods of stress.

The third point involves endocrine metrics. To monitor the long-term health of soldiers in order to catch early warning signs of adverse stress, it has become increasingly important to compare baseline hormone levels with changes attributable to intense stress. Measuring hormone concentrations currently requires the collection of fluids such as saliva or blood, and laboratory analysis lasting hours to days. The development of rapid fieldable hormone assays for use by medics in the field would be of great utility for tracking stress responsivity of soldiers. This could be in the form of a litmus paper-like assay, or a credit card sized lab-on-a-chip device, that would ideally respond to readily available fluid sources such as urine, saliva, or sweat. We believe that the development of a non-invasive hormone sensor sweat patch is possible in a 5-year time frame and would provide an inexpensive, light-weight means for tracking stress hormones in soldiers.

The final point is a necessary step for translating highly accurate laboratory measures to the imperfect conditions of the battlefield. Whereas distraction during tasks requiring concentration can be observed using fMRI, it will be necessary to identify a different neuro-monitoring device with greater field deployability. fMRI machines are too large and heavy, and require too much infrastructure and precise environmental controls to be useful in remote locations. As fMRI studies continue to increase our understanding of the neural basis of loss of vigilance, it will become possible to identify other neuroimaging devices that can be used as a proxy.

Accurately assessing an individual's stress reactivity profile would best be informed by data from functional brain activity, grey and white matter structure, baseline and event-responsive hormone levels, behavior, genetics, and further endpoints. These may not all be available or desirable measures in assessing individual stress response profiles in soldiers. Correlations between different measurements of neural function or stress should take into consideration the unique time frame of each measure. For example, whereas neuronal signaling in the limbic emotion-processing center of the brain can be very rapid (in the 1–200 msec range), hormone signaling in response to stress can endure for hours to days. Thus, if a stressful event such as an insurgent attack shortly precedes a peaceful activity like interaction with civilians, lingering high stress hormone levels may interfere with the Soldier's ability to function effectively.

Conclusions

Pre-deployment assessment measures of stress susceptibility and real-time operational status updates in the battlefield arena would contribute greatly to the early identification of cognitive difficulties in soldiers. These measures would furthermore improve scientific and clinical understanding of PTSD and soldier suicidal behaviors, perhaps even opening more avenues for intervention. Mission success, reduced morbidity/mortality, and prevention/treatment of PTSD would benefit from a greater understanding of the physiological stress response. For example, it may one day be possible to rapidly measure a soldier's cortisol levels immediately following a traumatic event to determine whether a threshold concentration is crossed, thereby putting him/her at risk for long-term psychological sequelae. Knowledge of pre-deployment baseline hormone levels will be necessary for comparison. Field-deployable neuroimaging devices could notify a commanding officer when a soldier should be relieved due to loss of concentration, or could directly alert the soldier to take the necessary measures to revive vigilance. This future capability will rely on further research into which brain area(s) and which neuroimaging devices provide the most accurate measures of loss of attentiveness. The Commander's Dashboard providing the commanding officer with a schematic approximation of soldier readiness is attainable using today's technologies and knowledge, although it should be thoroughly tested for utility and accuracy. In addition, it should be modifiable so that additional measures can be added in future iterations of the device.

The prediction of individual susceptibility to stress or fatigue will also be made possible in the near future. Currently, identification of resilient individuals occurs *post-hoc* of an extremely stressful experience. A valid goal for the 5-year timeframe would be to develop a suite of tests based on current knowledge, incorporating targeted genetic analysis, behavioral testing, and brain structural and functional scans that can provide an assimilated multimodal prediction of a soldier's stress reactivity during situation-specific activities. With the present state of scientific knowledge, we can begin to design tests to predict with some accuracy those who will be at great risk for stress-related loss of performance abilities, and those who are likely to excel.

Index